Cornelia Bäucker

Interactions of invasive plants with soil biota

Cornelia Bäucker

Interactions of invasive plants with soil biota

Do arbuscular mycorrhizal fungi contribute to the success of two non-native plant species?

Südwestdeutscher Verlag für Hochschulschriften

Impressum / Imprint
Bibliografische Information der Deutschen Nationalbibliothek: Die Deutsche Nationalbibliothek verzeichnet diese Publikation in der Deutschen Nationalbibliografie; detaillierte bibliografische Daten sind im Internet über http://dnb.d-nb.de abrufbar.
Alle in diesem Buch genannten Marken und Produktnamen unterliegen warenzeichen-, marken- oder patentrechtlichem Schutz bzw. sind Warenzeichen oder eingetragene Warenzeichen der jeweiligen Inhaber. Die Wiedergabe von Marken, Produktnamen, Gebrauchsnamen, Handelsnamen, Warenbezeichnungen u.s.w. in diesem Werk berechtigt auch ohne besondere Kennzeichnung nicht zu der Annahme, dass solche Namen im Sinne der Warenzeichen- und Markenschutzgesetzgebung als frei zu betrachten wären und daher von jedermann benutzt werden dürften.

Bibliographic information published by the Deutsche Nationalbibliothek: The Deutsche Nationalbibliothek lists this publication in the Deutsche Nationalbibliografie; detailed bibliographic data are available in the Internet at http://dnb.d-nb.de.
Any brand names and product names mentioned in this book are subject to trademark, brand or patent protection and are trademarks or registered trademarks of their respective holders. The use of brand names, product names, common names, trade names, product descriptions etc. even without a particular marking in this works is in no way to be construed to mean that such names may be regarded as unrestricted in respect of trademark and brand protection legislation and could thus be used by anyone.

Coverbild / Cover image: www.ingimage.com

Verlag / Publisher:
Südwestdeutscher Verlag für Hochschulschriften
ist ein Imprint der / is a trademark of
OmniScriptum GmbH & Co. KG
Heinrich-Böcking-Str. 6-8, 66121 Saarbrücken, Deutschland / Germany
Email: info@svh-verlag.de

Herstellung: siehe letzte Seite /
Printed at: see last page
ISBN: 978-3-8381-3592-2

Zugl. / Approved by: Berlin, FU, Diss., 2012

Copyright © 2013 OmniScriptum GmbH & Co. KG
Alle Rechte vorbehalten. / All rights reserved. Saarbrücken 2013

FOREWORD

This work was carried out between the years 2008 and 2012 under the supervision of Prof. Dr. Matthias C. Rillig at the Institute of Biology at the Freie Universität Berlin, Germany. This book is a cumulative work of three studies, which are referred by their Roman numerals.

I   Divergent responses of *Ambrosia artemisiifolia* to natural AM fungal communities in the new European range
II Non-native *Ambrosia artemisiifolia* are more influenced by relative density and identity of neighboring plant species than arbuscular mycorrhiza
III Distinct seed morphs of *Galinsoga parviflora* (Asteraceae) give rise to different soil feedbacks

I would like to thank Matthias Rillig for giving me the opportunity to do research into the exciting world of arbuscular mycorrhizal fungi. I am also deeply grateful to Kathryn Morris and Tancredi Caruso for discussions and statistical advice. I thank Jeff Powell and Edith Hammer for their comments on drafts, and Sabine Artelt and Sabine Buchert for their help during the setup of the target–challenger experiment, as well as Anika Lehmann and Weishuang Zheng for their help during the harvest of the final feedback experiment. Further, I thank the Dahlem Centre of Plant Sciences of Freie Universität Berlin for technical support and Michael Baumecker for the possibility to use soil from Thyrow.

TABLE OF CONTENTS

FOREWORD………………………………………………………………….. v

LIST OF TABLES……………………………………………………………... xi

LIST OF FIGURES……………………………………………………………. xiii

1   GENERAL INTRODUCTION…………………………………………….. 1

2   STUDY I: DIVERGENT RESPONSES OF *AMBROSIA ARTEMISIIFOLIA* TO NATURAL AM FUNGAL COMMUNITIES IN THE NEW EUROPEAN RANGE……….. 11

  2.1 Abstract…………………………………………………………….. 11
  2.2 Introduction………………………………………………………... 11
  2.3 Materials and methods…………………………………………….. 15
    2.3.1 Site characteristics…………………………………………. 15
    2.3.2 Soil sampling and analyses………………………………… 16
    2.3.3 Experiment………………………………………………….. 17
    2.3.4 Statistical analyses…………………………………………. 19
  2.4 Results……………………………………………………………... 20
    2.4.1 Soil properties……………………………………………… 20
    2.4.2 Variance explained by the first and second principal components…. 20
    2.4.3 Soil effects………………………………………………….. 21
    2.4.4 Plant responses to soil and mycorrhizal treatment………….. 21
    2.4.5 Plant origin–soil interaction and effects of plant origin……. 23
  2.5 Discussion…………………………………………………………. 24
    2.5.1 Mycorrhizal functions depending on soil and inoculum source……. 24
    2.5.2 Adaptation to soil environment and plant origin effects……. 26
    2.5.3 Conclusion………………………………………………….. 28
  2.6 References…………………………………………………………. 29
  2.7 Tables and figures…………………………………………………. 36

3 STUDY II: NON-NATIVE *AMBROSIA ARTEMISIIFOLIA* ARE MORE INFLUENCED BY RELATIVE DENSITY AND IDENTITY OF NEIGHBORING PLANT SPECIES THAN ARBUSCULAR MYCORRHIZA... 43

- 3.1 Abstract.. 43
- 3.2 Introduction.. 43
- 3.3 Materials and methods... 46
  - 3.3.1 Target–challenger experiment.. 47
  - 3.3.2 Pairwise competition experiment................................... 49
  - 3.3.3 Mycorrhizal colonization in the roots............................. 50
  - 3.3.4 Statistical analyses.. 50
- 3.4 Results.. 51
  - 3.4.1 Target–challenger experiment.. 51
  - 3.4.2 Pairwise competition experiment................................... 54
- 3.5 Discussion... 55
- 3.6 References... 61
- 3.7 Tables and figures.. 66

4 STUDY III: DISTINCT SEED MORPHS OF *GALINSOGA PARVIFLORA* (ASTERACEAE) GIVE RISE TO DIFFERENT SOIL FEEDBACKS....................... 73

- 4.1 Abstract.. 73
- 4.2 Introduction.. 74
- 4.3 Materials and methods... 77
  - 4.3.1 Study species.. 77
  - 4.3.2 Soils and soil preparation.. 78
  - 4.3.3 Experiment... 79
  - 4.3.4 Statistical analyses.. 81
- 4.4 Results.. 82
  - 4.4.1 Feedback contrasts of plants from the two seed types differ........... 83
  - 4.4.2 Effects of soil and the soil–soil treatment interaction.................. 84
  - 4.4.3 Effects of the seed history–seed type interaction........................ 84

4.5 Discussion..	85
4.6 References...	90
4.7 Tables and figures...	96

5   **SUMMARY**... 101
    5.1 Synthesis.. 106
    5.2 Future perspectives.. 108

6   **REFERENCES TO GENERAL INTRODUCTION AND SUMMARY**................ 109

APPENDIX A: Overview on existing theories mentioned in the book.............. 123
APPENDIX B: Supplemental Tables A.I.1–A.I.3 to Study I....................... 132
APPENDIX C: Supplemental Tables A.II.1–A.II.5 to Study II..................... 136
APPENDIX D: Supplemental Tables A.III.1–A.III.2 to Study III................... 139

LIST OF TABLES

Table I.1	Characteristics of the soils used in the experiment...................	36
Table I.2	Eigenvectors of the first five principal components of the Principal Component Analyses (PCAs).........................	36
Table I.3	Analyses of variance (ANOVAs) on the first principal component score (PC1) of the Principal Component Analyses.	38
Table II.1	Analyses of variance (ANOVAs) on response variables of *Ambrosia artemisiifolia* and neighboring plant species in the target–challenger experiment..	66
Table II.2	Analyses of variance (ANOVAs) on biomass traits of *Ambrosia artemisiifolia* and *Daucus carota* in the pairwise competition experiment..	67
Table II.3	Results of paired *t*-test on shoot and root biomass of *Ambrosia artemisiifolia* and *Daucus carota* in intra- and interspecific competition, and in the presence or absence of AM fungi.........	68
Table III.1	Mixed effect model analysis on responses of biomass and reproductive traits of *Galinsoga parviflora* in the feedback experiment..	96
Table III.2	Response variables of plants from the distinct seed types of *Galinsoga parviflora* to soil treatment in the experiment..........	97
Table III.3	Response variables of plants from the distinct seed types of *Galinsoga parviflora* to soil and soil treatment....................	97
Table A.1	Overview on existing theories mentioned in the book..............	123

Table A.I.1	Analyses of variance (ANOVAs) on the second principal component score (PC2) of the Principal Component Analyses.	132
Table A.I.2	Biomass traits, root traits and mycorrhization of *Ambrosia artemisiifolia* in response to soil and mycorrhizal treatment....	133
Table A.I.3	Biomass variables and root traits of *Ambrosia artemisiifolia* in response to soil and plant origin of *A. artemisiifolia*..............	135
Table A.II.1	Vegetative and reproductive variables of the target–challenger experiment in response to relative density of *Ambrosia artemisiifolia*, neighboring plant species and soil treatment.....	136
Table A.II.2	Response variables of the target–challenger experiment indicating a significant effect of neighboring species.............	137
Table A.II.3	Shoot biomass of *Ambrosia artemisiifolia* in response to its relative density and the four neighboring plant species............	137
Table A.II.4	Shoot biomass of *Ambrosia artemisiifolia* in response to neighboring plant species tested and soil treatment...............	138
Table A.II.5	Percentages colonization by AM fungal structures in roots of *Ambrosia artemisiifolia* and *Daucus carota* in the pairwise competition experiment..	138
Table A.III.1	Mixed effect model analysis on percentages of AM fungal structures and percentage root colonization by non-AM fungi in soils inoculated with trained soil.................................	139
Table A.III.2	Analysis of variance (ANOVA) on seed weight of seeds produced by *Galinsoga parviflora* during the second training round of the experiment...	140

LIST OF FIGURES

Figure I.1	Colonization with arbuscular mycorrhizal (AM) fungi in roots of *Ambrosia artemisiifolia* in the experiment...................….……..	38
Figure I.2	Loading plots of the Principal Component Analyses (PCAs)....	39
Figure I.3	Responses to soil and mycorrhizal treatments of *Ambrosia artemisiifolia* in the experiment..................…..………...	40
Figure I.4	First principal component score of the PCA on biomass traits of *Ambrosia artemisiifolia* indicating local adaptation to roadside soil in the experiment..	41
Figure II.1	Target versus challenger arrangement of *Ambrosia artemisiifolia* in the target–challenger experiment...............…...	69
Figure II.2	Shoot biomass of the four neighboring plant species in response to relative density of *Ambrosia artemisiifolia*.........	69
Figure II.3	Shoot biomass (a) and number of male inflorescences (b) of *Ambrosia artemisiifolia* in response to its relative density, soil treatment and the four neighboring plant species.....……………	70
Figure II.4	Shoot biomass of *Ambrosia artemisiifolia* and *Daucus carota* in response to the main effects soil treatment (a) and competitive situation (b)...	71
Figure III.1	Design of the feedback experiment with a soil training phase performed over two plant generations............................………	98
Figure III.2	'Trained vs. sterile' soil contrasts of plants grown from non-pappus and pappus seeds of *Galinsoga parviflora* in the experiment..................………...…………………………………	99

1 GENERAL INTRODUCTION

Earth is a dynamic system. Episodes of enormous interchange of species have been repeatedly taken place, for example in consequence of tectonic activity or during Pleistocene ice ages (e.g. Vermeij 1991; CLIMAP Project 1976; Brown and Sax 2004). Besides geographical rearrangements and climate changes, many species also fluctuate in their range expansions as a result of biological interactions on the timescale of decades to years, e.g. empirically observable as succession. The interchange of species, therefore, is a constantly occurring biological process that should not be viewed as abnormal *per se* (Vitousek 1992; Lodge 1993).

Since the last century, however, earth's biota is being homogenized rapidly as human activities increasingly introduce species outside their natural range (Elton 1958; Lodge 1993). After human-caused habitat destruction, biological invasions were exposed and afterwards heavily cited as the second largest component for current biodiversity loss (Vitousek 1992; Wilcove et al. 1998; Davis 2011). In fact, ecosystems worldwide face tremendous changes; the displacement of native species by non-native species has numerous direct and indirect effects on ecosystem functioning (D'Antonio and Vitousek 1992; Mack et al. 2000, Cassey et al. 2005; Wardle et al. 2011). Undoubtedly, the vast majority of habitat and community changes driven by the spread of non-native, invasive organisms must be regarded as irreversible (Sala et al. 2000). Furthermore, invasive species influence ecosystem services that are fundamental to human well-being resulting in substantial economic costs (see for USA: Pimentel et al. 2000, 2005; for South Africa, Hawaii, Great Lakes (USA): Pejchar and Mooney 2009; for Europe: Vilà et al. 2010; Keller et al. 2011; for China: Wan et al. 2010). Therefore, to understand why some species become the dominant component in communities, where they are not native, is of great scientific, economic and social interest as human activities such as international trade, transport and travel, which cause species dispersal into new ranges, continue to expand (Keller et al. 2011).

From the scientific perspective, biological invasions may also be regarded as 'grand, but unplanned, biological experiments' that give the opportunity to study ecological and evolutionary processes (Mooney and Cleland 2001; Brown and Sax 2004). During the last 50 years, patterns and mechanisms of biological invasions have been increasingly investigated (Richardson and Pyšek 2008); the number of articles published on invasive topics per year has been exponentially growing since the last 30 years (Kühn et al. 2011).

Terminology in invasion biology lacks uniformity and agreements (e.g. Richardson et al. 2000a; Davis and Thompson 2000; Daehler 2001; Colautti and MacIsaac 2004; Pyšek et al. 2004a; Inderjit 2005; Valéry et al. 2008; Colautti and Richardson 2009; Young and Larson 2011; Webber and Scott 2012). The existing terminological difficulties result from the fact that invasion research is multifaceted: the perspectives range from population or community ecology to evolutionary and molecular biology, as well as restoration. Moreover, the usage of terms is related to research histories and traditions. In the German-speaking part, for example, plants introduced before the discovery of America are termed 'archaeophytes', and those introduced after 1492 are 'neophytes' (e.g. Schroeder 1969; Kowarik 2003; Pyšek et al. 2004b). Recently, European scientists also suggested the term 'neobiota' as a value-neutral approach (Kowarik and Starfinger 2009; Kühn et al. 2012). *Sensu* Kowarik (2002) neobiota are organisms, independent of their taxonomic rank, that occur in a region beyond their native range due to human agency or that evolved from such taxa. This 'neobiota' concept, however, is not generally accepted.

The English terminology, in contrast, terms species outside their native ranges 'alien' (e.g. Crawley et al. 1996), 'imported' (e.g. Williamson and Fitter 1996a), 'non-indigenous' (e.g. Mack et al 2000; Pimentel et al. 2000), 'casual' or 'naturalized' (e.g. Richardson et al. 2000a), but also 'adventive' (e.g. Mühlenbach 1979) or 'non-native' (e.g. Davis et al. 2000). Additionally, more ambiguous vocabulary like 'exotic' (e.g. Keane and Crawley 2002) or 'invasive' is very common. Above all, the term 'invasive' is the major focus of the terminological debate inclusive related terms like 'invasion', 'invader', 'invasiveness' or 'invasibility' (e.g. Richardson et al. 2000a; Colautti and MacIsaac 2004; Richardson and Pyšek 2006; van Kleunen et al. 2010; Catford et al. 2012). Defining invasive species, some authors put emphasis on a large negative ecological and/or economical impact (Davis and Thompson 2002; Inderjit 2005), while others see the disruption of the local target community as the most important feature (Keane and Crawley 2002).

In the present book, I refer to the definition of an invasive plant species proposed by Richardson et al. (2000a). It states that invasive plants are 'naturalized plants that produce reproductive offspring, often in very large numbers, at considerable distances from parent plants and thus have the potential to spread over a considerable area'. Moreover, I see the invasion process as recently presented by Blackburn et al. (2011). Their concept can be regarded as a unifying framework; it incorporates classical concepts of Williamson (1996) and Richardson et al. (2000a), as well as other conceptual ideas (e.g. Heger and Trepl 2003; Colautti and MacIsaac 2004). The framework includes four stages, which are transport, introduction, establishment and spread. Moreover, it proposes that a species has to overcome a series of six barriers to become a fully invasive species. The barriers are: geography, captivity or cultivation, survival, reproduction, dispersal and environmental; hence, the framework makes no distinction between disturbed and undisturbed habitats (cf. Richardson et al. 2000a). Furthermore, the different possibilities of establishment (spontaneous and permanent) as proposed by Heger and Trepl (2003), are illustrated as a feedback loop between barriers to survival and reproduction in the unifying framework. Blackburn et al. (2011), however, do not rank first in conceptualizing a unifying approach for invasion ecology. Already 100 years ago, the Swiss botanist Albert Thellung suggested a universal framework (Kowarik and Pyšek 2012), and also other integrative concepts have been presented in the last few years (Barney and Whitlow 2008; Moles et al. 2008; Catford et al. 2009).

Aside from the passionate debate about terms and frameworks, invasion biology is a discipline that has been more characterized by theory accumulation than theory discrimination until recently (cf. Davis 2011); the number of theories explaining the exceptional success of invasive species is overwhelming. Some review articles, however, compose overviews on the existing leading theories (Sakai et al. 2001; Hierro et al. 2005; Mitchell et al. 2006; Catford et al. 2009). Catford et al. (2009) point out that many of the existing hypotheses are redundant as they 'overlap, mirror, unite or share similarity with pre-existing hypotheses'. For example, resource availability in the new environment is an integral component of several hypotheses, e.g. fluctuating resource availability (Davis et al. 2000), disturbance (Sher and Hyatt 1999), opportunity window (Shea and Chesson 2002), dynamic equilibrium model (Huston 2004), or environmental heterogeneity (Melbourne et al. 2007). The excessive generating of new hypotheses in the past may be viewed as resulting from the intention to find the 'holy grail' of invasion, as well as related to the fact that most studies focused on one single aspect of invasions only (Richardson and Pyšek 2008; Catford et al. 2009).

Every invasion process, however, must be understood as highly context-dependent and linked to a combination of both abiotic and biotic factors, and multiple mechanisms (e.g. Daehler 2003; Richardson and Pyšek 2006; Barney and Whitlow 2008). Therefore, the success of invasive species cannot be explained with mono-causality. Moreover, all factors and mechanisms underlying the rapid range expansion of invasive species must be assumed to vary in time and space. Recently, two studies demonstrated that even enemy release, which is a crucial aspect of many theories in invasion ecology, does not persist forever (Mitchell et al. 2010; Diez et al. 2010).

To categorize the different approaches and hypotheses that explain the mechanisms of biological invasions, Heger and Trepl (2003) emphasize four different approaches: (1) to focus on the characteristics of the invading species, (2) on those of the ecosystems invaded, (3) on the relationship between these two factors (key–lock approach), and (4) the invasion process in time. Catford et al. (2009) also synthesized four categories/factors, into which hypotheses might be divided: (1) human interference, (2) propagule pressure, (3) abiotic and (4) biotic factors. Recently, Jeschke et al. (2012) evaluated six of the major leading theories, which were classified by their main focus into three groups: (1) invaders themselves, (2) ecosystems into which the invaders were introduced, and (3) invader-ecosystem interaction. According to Jeschke et al. (2012), I also group theories based on their main focus in the present book. Here, I differentiate between the following three foci/categories that might be derived from hypotheses in invasion biology:

i) features of the invasive species
ii) characteristics of the new environment/habitat
iii) interactions of invasive species with their new environment

The first category 'features of the invasive species' refers to hypotheses like ideal weed (Elton 1958; Baker 1965, 1974) or propagule pressure (Williamson and Fitter 1996b; Lonsdale 1999). The theory of propagule pressure, which implies that the chance for successful invasion is increased by a high supply and frequency of plant propagule introductions, had been found to be a significant predictor of invasion in a meta-analysis (Colautti et al. 2006). Other theories belonging into this group primarily focusing on invasive species traits might be lag-phase (Kowarik 1995) or evolution of increased competitive ability (Blossey and Nötzold 1995), although the latter also has a strong interaction focus.

The second category 'characteristics of the new environment/habitat' addresses to hypotheses like fluctuating resource availability (Davis et al. 2000), disturbance (Sher and Hyatt 1999) or other theories explaining invasion success predominantly from the perspective of resource availability in the new environment (see above). This category is equivalent to the abiotic factor class suggested by Catford et al. (2009) and the aspect 'ecosystems into which the invaders were introduced' by Jeschke et al. (2012). Doubtless, abiotic characteristics play a major role for successful establishment and spread of invasive species. In a recent meta-analysis, globally widespread species were demonstrated to be better able to utilize increased resource amounts of nutrients, light and water compared to less widespread species (Dawson et al. 2012). Besides, Colautti et al. (2006) showed that disturbance and resource availability are significantly positively associated with invasibility. Furthermore, resource availability may synergistically interact with enemy release giving the advantage to non-native over native species (Blumenthal 2006; Blumenthal et al. 2009).

Among the proposed third category 'interactions of invasive species with their new environment' count hypotheses like enemy release (Keane and Crawley 2002), novel weapons (Callaway and Ridenour 2004), increased nitrogen cycling (Rout and Callaway 2009), enhanced mutualism (Reinhart and Callaway 2006), mycorrhizal degradation (Vogelsang et al. 2004), or invasional meltdown (Simberloff and Von Holle 1999). According to Jeschke et al. (2012), hypotheses considering invader–ecosystem interactions, such as enemy release, novel weapons, invasional meltdown are better supported than those, which exclusively focus on ecosystem properties or solely on invaders, like tens rule (Williamson and Brown 1986; Williamson and Fitter 1996a). Jeschke et al. (2012), moreover, report that the invasional meltdown theory has the highest level of support across both animals and plants in terrestrial, freshwater and marine habitats. The support, however, has considerably declined over time as it was found for all theories tested. For definition of all hypotheses mentioned in the book see Appendix A, Table A.1.

The aim of the present work is to make a contribution to the field of biotic interactions of invasive plants with soil biota. Soil biota include a wide range of taxa, for example mites, collembola, nematodes, macro-arthropods as beetle larvae, earthworms, enchytraeid worms, fungi like Glomeromycota, Basidiomycota, Ascomycota, as well as bacteria, and archaea. This highly diverse belowground community is known to drive aboveground community structure/functioning via direct and indirect pathways to plants (Wardle et al. 2004). Therefore, plant interactions with soil biota have been subject of numerous research projects in invasion ecology, and many studies found evidence for soil biota playing a crucial role in plant invasions, e.g. nematodes (van Ruijven et al. 2003; van der Putten et al. 2005), ectomycorrhizal fungi (Richardson et al. 1994; Wolfe et al. 2008; Nuñez et al. 2009; Trocha et al. 2012), arbuscular mycorrhizal fungi (e.g. Marler et al. 1999; Mummey and Rillig 2006; Stinson et al. 2006; Vogelsang and Bever 2009; Seifert et al. 2009), fungal pathogens (Mangla et al. 2008), fungal endophytes (Aschehoug et al. 2012), various N_2-fixing bacteria (e.g. Vitousek et al. 1987; Parker et al. 2006; Rout and Chrzanowski 2009), or soil microbes < 20 µm (e.g. Klironomos 2002).

Recently, Inderjit and van der Putten (2010) synthesized an overview on how soil biota may directly and indirectly interact with invasive plants, and point out that many questions are open because soil is often used as 'black box' in experiments. Indeed, studies using whole soil as treatment have a low mechanistic resolution, but make that compromise to gain greater realism. For example, in soil feedback studies it had been found that the magnitude of soil biota net effects on invasive plants is considerably less negative or even positive in new ranges compared to the species' native range (Reinhart et al. 2003; Callaway et al. 2004a; Reinhart and Callaway 2006; Kulmatiski et al. 2008; Callaway et al. 2011). Therefore, differences in soil feedback between 'home' and 'away' ranges may contribute to the successful spread of invasive plants, although these feedback differences must be assumed to become less important with increase in residence time. Diez et al. (2010) studied soil feedback responses of 12 non-native established plant species in New Zealand and found that those species that have been established longest (210 years) exhibited greater negative soil feedbacks than those with shorter residence time. The mechanisms behind theses feedback changes in 'away' ranges over time is not fully understood, but most probably related to accumulating plant–pathogen interactions, alike novel plant–herbivore interactions aboveground (Verhoeven et al. 2009). However, pathogens may switch very slowly from native to introduced species. As Mitchell et al. (2010) demonstrated long established plants (since more than 400 years) still had 60 % fewer pathogens in their new North American range compared to their native European range. Nonetheless, negative impacts of pathogens on non-native species must be assumed to accumulate over time. Aside from residence time, pathogen richness of non-native plants also may depend on geographic size of the introduced range, and if plants have a history of agricultural use (Mitchell et al. 2010).

This book primarily focuses on the impact of arbuscular mycorrhizal (AM) fungi on invasive plants. As non-native species, I studied *Ambrosia artemisiifolia* L. and *Galinsoga parviflora* Cav. in the new European range. Both plant species are annual and belong to the family of Asteraceae, whose members often form mycorrhizal symbioses with AM fungi, phylum Glomeromycota (Schüßler et al. 2001). The AM symbiosis is in most cases facultative for the plant partner, but always obligatory for the fungus (Helgason and Fitter 2009). The association most probably evolved in the Ordovician; fossil records date back to 460 million years ago (Redecker et al. 2000). AM fungi provide nutrients, predominantly phosphorous (P) to the plant side in exchange for carbohydrates (Allen 1991; Smith and Read 2008). In addition to the reciprocal nutrient fluxes, AM fungi mediate other functions to plants, such as pathogen protection (e.g. Newsham et al. 1995; Borowicz 2001; Wehner et al. 2010; Veresoglou and Rillig 2011) or improved resistance against drought stress (Augé 2001; Augé et al. 2004). Moreover, the mycorrhizal status of a plant alters their competitive ability, which maintains plant diversity (van der Heijden et al. 1998; Hart et al. 2003). Furthermore, multitrophic interactions as leaf-mining herbivores with parasitoids differ depending on mycorrhization of plants (Gange et al. 2003). However, under certain environmental conditions, such as high nutrient availability in fertilized systems or reduced photosynthesis, AM fungi seem to be less cooperative to the plant side resulting in reduced plant performance (Kiers et al. 2011). Therefore, the AM fungal symbiosis was viewed to act in a range of functions from mutualism to parasitism mediated by abiotic and biotic environmental conditions, including the plant and fungal genotype (Johnson et al. 1997).

Regarding the success of invasive plants, AM fungi have been found to be of particular importance in a number of cases. There are some hypotheses which highlight plant interactions with AM fungi as the critical factor for the plant's invasive spread: enhanced mutualism (Reinhart and Callaway 2006) and mycorrhizal degradation (Vogelsang et al. 2004) (for details of the theories see Appendix A, Table A.1). A few review articles, moreover, cover the topic of plant interaction with AM fungi in the context of invasion (Richardson et al. 2000b; Wolfe and Klironomos 2005; Mitchell et al. 2006; Pringle et al. 2009; Shah et al. 2009). Recently, Moora et al. (2011) showed that the palm *Trachycarpus fortunei* associates with widely distributed AM fungal taxa when it was introduced to different new European ranges; hence, non-native mycorrhizal plants select for AM fungal generalists and, therefore, seem to be not limited by a lack of mutualistic fungi (Richardson et al. 2000b). Consequently, the role of AM fungi has to be taken into account to properly explain invasive success of plant species (Mitchell et al. 2006). The present work, therefore, considers questions of invasive plants and natural arbuscular mycorrhizal (AM) communities.

I investigated three different issues, which correspond to the studies I–III. Because AM fungal taxa and isolates have been shown to differ in their functions (e.g. van der Heijden et al. 1998; Bray et al. 2003; Munkvold et al. 2004; Scheublin et al. 2007), I aimed to conduct my experiments with comparatively high realism. In my studies, therefore, I always maintained the ecological context of soil and natural AM fungal communities.

Study I reports about the relationship of the non-native plant *Ambrosia artemisiifolia* with natural AM fungal communities in the new European range at local scale. In a reciprocal inoculation experiment, I studied whether or not the mycorrhizal symbiosis between *A. artemisiifolia* and native AM fungal communities shows evidence of co-adaptation in a European roadside and cornfield population, respectively. I expected that plant performance and fitness of *A. artemisiifolia* is greater when the plants, soil and AM fungal community come from the same site. Further, I predicted that the AM fungal community from the roadside habitat would act cooperatively, while the AM fungal community from the agricultural field would show less cooperative behavior in its agricultural soil context.

Study II reports about the influence of natural AM fungal communities on the competitive ability of *A. artemisiifolia*. The performance of *A. artemisiifolia* was studied in two different relative abundances (target and challenger arrangements), as well as in the presence of different neighboring plant species of the European range. As neighboring plants, I tested *Conyza canadensis* L., *Artemisia vulgaris* L., *Daucus carota* L. and *Tanacetum vulgare* L., which I found co-existing with *A. artemisiifolia*. I expected that the mycorrhizal symbiosis would enhance the competitive ability of *A. artemisiifolia*; its invasive spread has been suggested to be promoted by AM fungi (Fumanal et al. 2006). Moreover, I investigated the influence of natural AM fungal communities on *A. artemisiifolia* and *D. carota* grown in pairwise arrangements of intra- and interspecific competition.

Study III focuses on the aspect of heterocarpy of the non-native plant *Galinsoga parviflora* Cav. in a soil feedback study. *G. parviflora* produces two distinct seed morphs: seeds equipped with a pappus for long-distance dispersal and non-pappus seeds for maintaining the existing population. I asked if the different dispersal capacities of the two seed types might correlate to different soil feedback responses, which may contribute to the successful spread of *G. parviflora* in the new range. Therefore, I tested soil feedback responses of plants grown from the two seed types (non-pappus and pappus seeds) in soil trained over two plant generations by *G. parviflora*. Considering the different dispersal abilities of the two seed types, I hypothesized that plants arising from non-pappus seeds would exhibit better performance, i.e. less negative soil feedback, in soil trained by the mother plant than those grown from pappus seeds.

2 STUDY I: DIVERGENT RESPONSES OF *AMBROSIA ARTEMISIIFOLIA* TO NATURAL AM FUNGAL COMMUNITIES IN THE NEW EUROPEAN RANGE

2.1 Abstract

Recently, existence of coadapted plant-arbuscular mycorrhizal (AM) fungal interactions has been found, but knowledge about the extent to which such adaptations also occur during plant invasions is lacking. Here, I investigated whether or not the mycorrhizal symbiosis between *Ambrosia artemisiifolia* and natural AM fungal communities shows evidence of co-adaptation in the new European range. In a reciprocal inoculation experiment with 'full soil strength' inocula, I compared performance of genotypes from two different sites: a roadside and a cornfield habitat. Natural AM fungal assemblages were mutualistic with *A. artemisiifolia* in roadside soil, but not in agricultural soil tested. Decreased plant growth in response to the less cooperative quality of the agricultural AM fungal community in the agricultural soil coincided with alterations of plant root systems towards greater fineness. I found no evidence for locally adapted plant-AM fungal interactions, but adaptation of roadside genotypes to a roadside soil environment. My results highlight the importance of the soil context for mycorrhizal functions. Contrasting effects of natural AM fungal communities and processes of adaptation to novel soil conditions may play a crucial role in the early stages of the spread of non-native *A. artemisiifolia*.

2.2 Introduction

Arbuscular mycorrhizal (AM) fungi, Phylum Glomeromycota, colonize plant roots gaining photosynthetically fixed carbon in exchange for mineral nutrients, predominantly phosphorus (e.g. Allen 1991; Pearson and Jakobsen 1993; Smith and Read 2008). In addition to the well-known reciprocal nutrient fluxes, AM fungi mediate multiple functions affecting plant traits (e.g. Newsham et al. 1995). For example, mycorrhizal plants are better defended against fungal pathogens (e.g. Borowicz 2001; Veresoglou and Rillig 2011), have greater reproductive output (Lu and Koide 1994; Koide and Dickie 2002; but see Allison 2002), and flower earlier (Sun et al. 2008). Furthermore, several studies report that plant species and genotypes with greater mycorrhizal responsiveness have root systems with a coarser root architecture than non-responsive species and genotypes (Hetrick 1991; Schultz et al. 2001; Berta et al. 2002; Seifert et al. 2009). The multidimensionality of plant responses to mycorrhization in part reflects the functional diversity of AM fungal taxa and isolates (e.g. van der Heijden et al. 1998; Klironomos 2003; Munkvold et al. 2004; Antunes et al. 2011).

In nature, plant species forming arbuscular mycorrhizas are typically colonized by a community of co-occurring AM fungal taxa, which form a complex underground mycelial network (Smith and Read 2008). There is accumulating evidence that plants interacting with this network are able to distinguish between cooperative and less cooperative AM fungi and promote more cooperative fungal partners with increased photosynthate allocation (Bever et al. 2009; Kiers et al. 2011). Therefore, selection pressure in arbuscular mycorrhizas would favor plant-fungi combinations which are most advantageous to both sides under the respective environmental factors (Helgason et al. 2002; Helgason and Fitter 2009; Johnson 2010). As a result of that coevolutionary selection process, both partners would specialize in their interactions and become locally adapted to each other and their abiotic environment (Thompson 2005; Hoeksema 2010).

Mycorrhizas might be understood as a dynamic 'coadapted mycorrhiza–soil complex' (Johnson et al. 1993), where both plant and fungal communities continuously adjust to the soil conditions and to one another. Therefore, if local circumstances allow a balanced trading partnership over time, co-adaptation in mycorrhizas and their local soil environment should be promoted (Johnson 2010). In this process, however, selection pressure may be stronger under extreme habitat conditions, for example in phosphorus and/or nitrogen limited soils, or if the interaction between native AM fungi and plant communities experienced abrupt changes due to introduction and spread of an invasive plant (Pringle et al. 2009; Richardson et al. 2000) altering the mycorrhizal interactions of the resident plant species (e.g. Marler et al. 1999; Mummey and Rillig 2006; Vogelsang and Bever 2009; Zhang et al. 2010).

AM fungal associations are not always advantageous for the plant (Johnson et al. 1993). Under certain environmental conditions, such as high nutrient availability in fertilized systems or reduced photosynthesis, plant growth is decreased by AM fungi (Johnson et al. 1997; Verbruggen and Kiers 2010). Consequently, the plant–AM fungi interaction is viewed to act in a range of functions from mutualism to parasitism mediated by abiotic and biotic environmental conditions, including the plant and fungal genotype (Johnson et al. 1997). Hence, abiotic or biotic variables resulting in disadvantageous effects on one side of the plant–AM fungus relationship might counteract local adaptation in mycorrhizas.

Our understanding of the extent to which local adaptation might be important in arbuscular mycorrhizas is still limited. To date, only two studies focused explicitly on that question and used the approach of reciprocal inoculation of natural AM fungal communities (Johnson et al. 2010; Ji et al. 2010). Johnson et al. (2010) found evidence for co-adaptation because complete 'home' combinations of soil, whole soil inoculum and plant ecotype of *Andropogon gerardii* resulted in the highest fitness of the symbiotic partners in all populations tested. The experiment, moreover, showed that AM fungi and other soil organisms sharing a history in a nitrogen-limited soil were more effective in nitrogen supply to the plant; they were hence locally adapted to soil conditions. In contrast, Ji et al. (2010) demonstrated that adaptation of plants with AM fungal communities might depend on the plant species. Plant growth of *Sorghastrum nutans* was increased when soils were inoculated with the respective 'local' AM fungal spore community, while the origin of the inoculum had no effect on *Schizachyrium scoparium*. Further, the taxonomic composition of the AM fungal spore communities was also reported to change when the fungal spore inocula were introduced to novel soils. Other studies not employing reciprocal transplanting of AM fungi between study systems also indicated that plants and AM fungi may be adapted in their interactions (Schultz et al. 2001; Klironomos 2003; Pánková et al. 2008; Seifert et al. 2009). For example, Schultz et al. (2001) found a greater growth response by ecotypes of *Andropogon gerardii* to AM inoculation in phosphorus-limited soil when plants came from these nutrient poor conditions. Klironomos (2003), testing single AM fungal isolate–plant interactions, showed that plant performance was more strongly affected, both positively and negatively, in 'home' combinations of plants and AM fungi compared to pairs where either the plant or the AM fungus were exotic. The equivocal results reported to date highlight the need for additional studies in different ecosystems.

Here I study the interaction between two natural AM fungal assemblages and *Ambrosia artemisiifolia* L. (Asteraceae) in the plant's new European range. I ask if coevolutionary dynamics may have already led to a coadapted mycorrhiza–soil complex within the introduced range, since there is a potential for coevolution to drive rapid and far-reaching change (Thompson 1999). Thus, I was interested in testing rapid evolution of local adaptation, which has been suggested as an important mechanism and a fundamental issue in invasion ecology (e.g. Sakai et al. 2001, Colautti et al. 2009). Recently, Buswell et al. (2011) showed that rapid adaptive evolution of introduced species might be more common and of greater importance than previously thought.

The study focus was the regional scale. Methodically, I used the approach of making comparisons between demes within habitats, which corresponds to the 'local vs. foreign' criterion (Kawecki and Ebert 2004). The 'local vs. foreign' contrast addresses the efficacy of divergent selection relative to other evolutionary processes, and has been proposed as diagnostic for the pattern of local adaptation (Kawecki and Ebert 2004). Therefore, I refer to the 'local vs. foreign' terminology; it has analogously already been used by Ji et al. (2010) studying local adaptation in mycorrhizas in a two-site comparison.

I addressed my question using *A. artemisiifolia* because the plant's invasive spread is thought to be facilitated by AM fungi (Fumanal et al. 2006). The species, moreover, is known to respond positively to mycorrhizal inoculation (Crowell and Boerner 1988). Further, the plant is ideal to quantify resource allocation to sexes because male and female functions are located in different types of flowers on each individual (e.g. Ackerly and Jasieński 1990; Friedman and Barrett 2011).

In a reciprocal inoculation experiment, I compared the performance of plant populations from two different sites: a roadside habitat and an agricultural field. The hypothesis was that plant performance and plant fitness are greater when plants, soil and AM fungal community come from the same site (complete 'local' combinations of plant origin, soil and AM fungi) compared to combinations including plants from the other site (here defined as 'foreign' plant origin). Further, I ask how the respective AM fungal assemblages function in their own (defined as 'local') soil vs. when they are introduced to new (defined as 'foreign') soils. For testing AM fungi in their 'local' soil context, I predicted that the AM fungal community from the roadside habitat would act cooperatively, while the AM fungal community from the agricultural field would show less cooperative behavior.

2.3 Materials and methods

2.3.1 Site characteristics

In October 2008, seeds of *Ambrosia artemisiifolia* were collected from individual plants from two sites in southern Brandenburg, Germany. At each location, plants selected for collecting seeds were randomly chosen and distributed across a wide area (more than 3000 m^2) of the population. The first site was a roadside habitat, where *A. artemisiifolia* plants form dense stands growing over a length of 0.5 km on either side of the road (51°44'02.20"N, 14°27'27.22"E). The second site was an agricultural field planted to *Zea mays* in the year of seed collection (51°45'15.20"N, 13°58'21.88"E). The distance between the sites was approximately 40 km. For both populations the year of introduction of *A. artemisiifolia* is unknown, but the region is found to be one large center of the plant's distribution in Germany (Brandes and Nitzsche 2006). In this area *A. artemisiifolia* occurs at the edges of cornfields, on fallows and stubble fields, in intercrop areas and on roadsides, typical of an uneven and disconnected distribution (Brandes and Nitzsche 2006).

Mycorrhizal fungal communities of the two sites exhibited obvious visual differences in terms of root colonization. Root colonization in the roadside habitat was dominated by a special group of AM fungi, which are described as '*Glomus tenue*' or fine endophytes (Thippayarugs et al. 1999): hereafter referred to as FE. FE are typical for acid soils and characterized by hyphal diameters of less than 1.5 µm (Figure I.1a). In contrast, roots from the agricultural field harbored predominately the 'normal' AM fungi (Figure I.1b): hereafter termed coarse AM fungi. In October 2008, roots of *A. artemisiifolia* of 10 individuals from each site were stained and analyzed. For the roadside habitat we found colonization levels of 53 ± 6 % for coarse AM fungi and 20 ± 5 % for FE. In the cornfield soil, roots were less colonized and showed colonization levels of 17 ± 6 % for coarse AM fungi and 6 ± 3 % for FE.

2.3.2 Soil sampling and analyses

In March 2009, soil was collected from the two locations. I took soil samples only from areas that were occupied by *A. artemisiifolia* during the preceding autumn. At each habitat six soil samples of 10 L each were taken from the top 12 cm of soil. Soil samples per site were pooled, mixed and sieved (5 mm). Half of the soil from each habitat was pasteurized by steaming (Sterilo 1K, Harter Elektotechnik, Schenkenzell, Germany) for four hours at 90 °C. The other half of the soil was used for inoculum extraction.

To analyze soil I took soil samples from each soil after the steaming process. The samples were air-dried, sieved through a 2 mm sieve and analyzed for pH, water repellency, mineral nitrogen (N), mineral carbon (C) and plant available phosphorus (P). Soil pH was determined using both deionized water and in a 1:3 soil:0.01 M $CaCl_2$ suspension (van Lierop and MacKenzie 1977). Water repellency was measured as the water drop penetration time (Doerr 1998). Mineral N and C contents were determined using a CN analyzer (EuroEA3000-Single), and plant available P was analyzed as calcium-acetate-lactate soluble phosphorus content according to the German standard method DIN 3.4.1.30.2a (Blume et al. 2000).

2.3.3 Experiment

The experiment had a fully 2 x 2 x 3 factorial design and was replicated 12 times. It consisted of all combinations of seeds of *A. artemisiifolia* from the two habitats (seeds collected from the roadside: hereafter roadside seeds, seeds collected from the cornfield: hereafter cornfield seeds), soil from those two locations (roadside soil, cornfield soil), and three soil treatments with two types of inocula (mycorrhizal community from roadside soil, mycorrhizal community from cornfield soil, non-mycorrhizal control).

Mycorrhizal inocula are often prepared from a much smaller volume of soil than that used to fill pots in experiments, leading to potentially unrealistic inocula soil:experiment soil ratios of 1:10 or 1:100. I conducted the experiment with mycorrhizal fungal inocula at the more realistic 'full soil strength', meaning that I extracted inocula from the same volume of soil used to fill pots. Soils were wet sieved and inocula were prepared as filtrate (38–212 µm). Our approach of adding inocula as filtrate including AM fungal communities and other soil organisms was different from Ji et al. (2010) using AM fungal spores, but comparable to Johnson et al. (2010), who added whole-soil inoculum. We chose this method to allow mycorrhizal colonization to establish starting from both AM fungal spores and hyphae (Klironomos and Hart 2002). Further, I checked the inocula for nematodes, which were not detected (binocular microscope). The control treatment received only a mixed microbial wash containing equal parts microbial wash (i.e. filtrate passing a 20 µm sieve) prepared from roadside and field soils (Koide and Li 1989). All treatment pots also received 15 mL of mixed microbial wash to correct for effects resulting from non-AM fungal microbial communities.

The seedlings were germinated in sterilized sand in the greenhouse (day temperature 25 °C, night temperature 16 °C, photoperiod 16 h). Before transplanting, 6 x 25 cm, 400 ml Conetainer pots (Stuewe and Sons., Oregon, USA) were filled with steam pasteurized roadside or cornfield soil. Then, 15 mL of a mixed microbial fraction was added to each pot, and those intended for inoculation received 15 mL of roadside or cornfield AM inoculum while non-inoculated controls received an equivalent amount of water. Plants grew in a fully randomized arrangement for ten weeks in a climate chamber (day temperature 20 °C, night temperature 18 °C, humidity 60 %). I chose a 14 hour photoperiod to encourage optimal development of *A. artemisiifolia* (Deen et al. 1998). Plants were watered as needed with the same amount of deionized water. Flowering status of the male inflorescences and number of seeds set were recorded weekly during the duration of the experiment.

At harvest, reproductive biomass was carefully removed from shoots. Roots were separated from soil and washed under a stream of water. Biomasses of roots, shoots and male inflorescences were determined after drying for four days at 40 °C. Both ripe and immature seeds were counted and weighed after air-drying to determine reproductive output.

To measure mycorrhizal root colonization, a root sample of each plant (ca. 100 mg of dry root material) was stained using the ink-vinegar method of Vierheilig et al. (1998). Mycorrhizal colonization levels were determined using the magnified intersect method of McGonigle et al. (1990) based on 100 intersections per root sample examined at 200X. Broad shifts in the composition of AM fungal communities in response to 'local' vs. 'foreign' soil were analyzed by assessing morphological structures of coarse AM fungi and FE (hyphae, arbuscules, vesicles) separately. To determine differences in root architecture, roots of 11 genotypes (six from the roadside, five from the field) were scanned (STD4800 Scanner, resolution 400 dpi) using WinRhizo image analysis system (version Pro 2007b, Régent Instruments Canada Inc.). Roots with root diameters < 200 µm were classified as fine roots and those between 200–1000 µm as coarse roots.

2.3.4 Statistical analyses

In local adaptation studies plant performance is commonly based on more than one individual trait, leading to non-independent data per plant individual (Kawecki and Ebert 2004); I thus statistically analyzed plant responses as a plant trait syndrome. Therefore, influence of the main effects (seed origin, soil, soil treatment) and their interactions on plant performance were investigated using Principal Component Analysis (PCA) on the correlation matrix (Legendre and Legendre 1998; Gotelli and Ellison 2004). This ordination approach enabled me to reduce the dimensionality of the multivariate dataset to a few non-correlated new variables. I performed three PCAs focusing on three different plant trait aspects: biomass, root traits and mycorrhization. Each single PCA included a set of seven variables.

PCA on plant biomass was calculated on standardized data of shoot and root biomass, mass of male inflorescences and both ripe and immature seeds, number of seeds produced after five weeks of growth, and total seed number. To meet assumptions of normality and homogeneity data concerning seed number were log-transformed, whereas all other biomass variables were Box-Cox transformed. Time of flowering was not included in the analysis because of missing values. PCA on root traits was computed from standardized data of variables of root length per volume, root surface area, both fine and coarse root volume, both fine and coarse root length, and root diameter. Here, measures of root diameter were log-transformed to meet assumptions of normality. PCA on mycorrhization included standardized variables of percentage total AM fungal root colonization, and percentage root colonization by hyphae, arbuscules and vesicles of both coarse AM fungi and FE. The original data were log-transformed to achieve normality.

To test for differences among treatment groups, principal component scores from the first and second axis of the PCA were analogously treated to univariate response variables, and analyzed by Analysis of Variance (three-way ANOVA). Differences between treatment groups were compared with Tukey HSD post-hoc comparison tests ($P < 0.05$). To identify differences in soil properties I used a two sample *t*-test ($P < 0.05$). All statistical analyses were performed using R version 2.10.1 (R Development Core Team 2009).

2.4 Results

2.4.1 Soil properties

The soils of the locations differed significantly in soil pH, N, and water repellency (Table I.1). Extractable contents of C and P were similar.

2.4.2 Variance explained by the first and second principal components

The first principal component (PC1) of the three PCAs on the different plant traits – plant biomass, root morphology, mycorrhization – of *A. artemisiifolia* accounted for more than 50 % of the variance (Table I.2, Figure I.2). In the PCA on root traits, variance explained by PC1 was even higher (70 %; see Table I.2). In all PCAs, a significant interaction term indicated that responses of PC1 scores to soil were mediated by mycorrhizal treatment (Table I.3). Moreover, for the PCAs on plant biomass and root traits I found that PC1 scores were also significantly influenced by the interaction between plant origin and soil. This effect, however, could not be shown for root traits with pairwise comparison tests (Tukey HSD; $P > 0.05$). Furthermore, PC1 scores were consistently strongly influenced by the factor soil (Table I.3).

Regarding the second principal component (PC2), patterns differed depending on the trait aspect analyzed. Details on analyses of variance (ANOVAs) on PC2 are reported in the Supplemental Table A.I.1. PC2 of the PCA on biomass explained 20 % of data variance (Table I.2), but PC2 scores were not significantly influenced by the factors tested (Supplemental Table A.I.1). PC2 of the PCA on root traits accounted for 25 % of the variance (Table I.2) and was significantly influenced by plant origin only (Supplemental Table A.I.1). In contrast, PC2 of the PCA on mycorrhization explained 42 % of the variance (Table I.2) and was significantly impacted by both the soil–

mycorrhizal treatment and plant origin–mycorrhizal treatment interactions (Supplemental Table A.I.1). Effects of the plant origin–mycorrhizal treatment term could not be shown with pairwise comparison tests (Tukey HSD; $P > 0.05$).

2.4.3 Soil effects

Soil conditions had a strong effect on the plant traits tested. Plants growing in roadside soil had considerably less vegetative biomass (202.8 ± 11.5 mg) compared to those in cornfield soil (665.9 ± 25.9 mg). In addition, seed mass of ripe seeds was on average more than doubled in cornfield soil (94.9 ± 9.0 mg) compared to roadside soil (44.4 ± 3.9 mg). Male flower biomass was also higher in cornfield soil (77.5 ± 3.9 mg) compared to roadside soil (28.9 ± 2.0 mg). Flowering started significantly earlier in cornfield soil (roadside soil: 4.5 ± 0.1 weeks, cornfield soil: 3.7 ± 0.1 weeks, soil effect: $F_{1,130} = 38.4$; $P < 0.001$, ANOVA).

Concerning root traits, plants growing in cornfield soil had twice the coarse root length (2310.3 ± 91.9 cm) compared to those in roadside soil (1149.6 ± 90.1 cm). Fine root length, moreover, was also higher in cornfield soil (640.7 ± 41.3 cm) than in roadside soil (490.1 ± 47.9 cm), although the increase was proportionally smaller. Hence, plants in roadside soil had finer root systems with on average smaller root diameters (0.180 ± 0.004 mm) compared to those in cornfield soil (0.250 ± 0.006 mm).

In terms of mycorrhization, plants grown in soils with AM fungal assemblages both from the roadside and the cornfield soil were colonized with AM fungi (Supplemental Table A.I.2). Non-mycorrhizal controls, however, were rarely infected by fungi (0.30 ± 0.10 %), none of which could be classified as AM fungi. Percentage of total AM fungal root colonization was on average 31 % higher in roadside soil (78 ± 2 %) than in cornfield soil (47 ± 3 %).

2.4.4 Plant responses to soil and mycorrhizal treatment

The vast majority of response variables indicated that effects of mycorrhizal treatments ('local', 'foreign' soil inoculum, and non-mycorrhizal control) on plant traits differed depending on the soil type. Details on measured variables in response to soil and mycorrhizal treatment are reported in the Supplemental Table A.I.2.

Regarding the soil–mycorrhizal treatment effect for both the PCAs on plant biomass and root traits, PC1 scores were negative in roadside soil; in cornfield soil the values were positive (Figures I.3a and I.3b). For the PCA on biomass, higher or less negative PC1 scores should be interpreted as greater biomass, because both shoot and root biomass, and also number of seeds had the highest loading on PC1 (Table I.2). In terms of root traits, higher PC1 scores reflect bigger and also coarser root systems, since variables of root surface area, root length per volume, and coarse root length had the highest influence on PC1 (Table I.2).

For plant biomass in roadside soil, non-mycorrhizal control plants had the lowest PC1 scores; hence, had significantly less biomass compared to other combinations tested. For 'local' and 'foreign' inoculum in roadside soil, PC1 scores were significantly higher compared to control, thus plant biomass was significantly increased in the presence of both AM fungal inocula (Figure I.3a). In cornfield soil, conversely, control plants had the highest PC1 scores, i.e. plants had greatest biomass in the absence of AM fungi. Moreover, 'local' soil inoculum caused a significant decrease in plant biomass compared to non-mycorrhizal controls in cornfield soil. Biomass of plants growing in cornfield soil with the 'foreign' inoculum was not different from those in control or 'local' inoculum treatments (Figure I.3a).

Regarding response patterns on root traits, PC1 scores did not significantly differ for the different mycorrhizal treatments in roadside soil (Figure I.3b). In cornfield soil, however, combination of cornfield soil with cornfield inoculum generated a significant decrease in PC1 scores compared to combinations with roadside inoculum (Figure I.3b). This means that in cornfield soil root systems were significantly smaller and less coarse when plants were treated with the 'local' compared to 'foreign' soil inoculum.

Concerning mycorrhization, we found strong proportional shifts between coarse AM fungi and FE colonizing roots among all treatment combinations of soil and mycorrhizal inoculum, which is indicated by both PC1 and PC2 scores revealing significant soil–mycorrhizal treatment interactions (Table I.3, PC2 soil x mycorrhizal treatment: $F_{1,132} = 16.41$; $P < 0.001$, ANOVA). In spite of these shifts, cornfield inoculum always resulted in higher root colonization by coarse AM fungi than FE; for the roadside inoculum higher proportions of FE structures were typical (Supplemental Table A.I.2). Differences in root colonization resulting from inoculum application to 'local' or 'foreign' soils were mainly related to the dominant AM fungal component of the respective inoculum only. Treating roadside soil with cornfield inoculum increased root colonization by coarse AM fungal structures, while for cornfield soil with roadside inoculum a strong decrease in FE colonization was found (Supplemental Table A.I.2).

Focusing on PC1, scores were more negative the higher the root colonization with FE and total AM fungi; non-mycorrhizal controls had the highest scores (see Figure I.3c). Thus, PC1 scores were lowest for the combination of roadside soil with roadside inoculum, which had highest colonization both with total AM fungi and FE (Figure I.3c, Supplemental Table A.I.2).

2.4.5 Plant origin–soil interaction and effects of plant origin

Regarding biomass patterns, PC1 scores significantly differed for 'local' and 'foreign' plant origin tested in roadside soil, with roadside origin plants being larger (Figure I.4). This indicates that in roadside soil plants with roadside origin produced significantly more biomass than plants originating from cornfield seeds (higher PC1 scores correlate with greater biomass, Figure I.2a).

In addition, function of the male gender was also significantly affected by the plant origin–soil interaction (plant origin x soil: $F_{1,132} = 11.82$, $P < 0.001$, ANOVA). Here, the difference was significant for cornfield soil only: plants with cornfield origin had significantly greater male flower biomass compared to plants with roadside origin (roadside origin: 66.0 ± 4.1 mg, cornfield origin: 89.3 ± 6.0 mg, $P = 0.01$ for pairwise comparison between roadside and cornfield plant origin in cornfield soil). Details on other variables in response to soil and plant origin are reported in the Supplemental Table A.I.3.

The factor plant origin significantly influenced PC2 scores of the PCA on root traits (plant origin: $F_{1,54}$ =9.52; P = 0.003, ANOVA). Because response variables describing fine root attributes had the highest loading on PC2 (Table I.2), I highlight fine root length and average root diameter. Plants originating from roadside seeds had on average smaller root diameters (0.209 ± 0.008 mm) compared to those from cornfield seeds (0.221 ± 0.007 mm). In addition, genotypes with roadside origin produced on average 43 % more fine root length (655.3 ± 42.3 cm) compared to those with cornfield origin (457.6 ± 44.4 cm). Plant origin also significantly influenced time of flowering (plant origin: $F_{1,130}$ = 12.5, $P < 0.001$, ANOVA). Plants originating from cornfield seeds flowered earlier (3.9 ± 0.1 weeks) than those from roadside seeds (4.3 ± 0.1 weeks).

2.5 Discussion

2.5.1 Mycorrhizal functions depending on soil and inoculum source

In this study, natural AM fungal assemblages were found to exhibit different qualities of cooperation with non-native *A. artemisiifolia* depending on mycorrhizal inoculum source and soil conditions. In less fertile roadside soil, presence of AM fungal communities significantly improved plant performance compared to the non-mycorrhizal control regardless of whether the inoculum came from the roadside or the cornfield habitat. The positive effect of mycorrhizal inoculation on *A. artemisiifolia* in roadside soil was reflected both in greater shoot and root biomass, and also higher number of seeds produced in comparison to the non-mycorrhizal treatment. As a result, enhanced vegetative plant growth was positively correlated with increased reproductive output indicating higher plant fitness: such a relationship is common in herbaceous plants (Allison 2002). In the comparatively adverse roadside habitat, which can also be expected to be corridors of invasion (Joly et al. 2011), mycorrhizal association may thus increase fitness of *A. artemisiifolia*, which may promote the plant's spread as suggested by Fumanal et al. (2006).

In contrast, in cornfield soil, where plants on average had almost triple the biomass compared to the roadside soils, we found no evidence of a beneficial role of AM fungi on plant performance. Consistent with empirical and theoretical studies (e.g. Johnson et al. 1993; Johnson et al. 1997; Verbruggen and Kiers 2010), our results demonstrate that the AM fungal community from the managed agricultural system acted less cooperatively, i.e. decreased plant biomass, but only in its 'local' soil. Thus, less cooperative or 'cheating' behavior of the cornfield AM fungal community was only present when the ecological soil–AM fungi context was maintained. The importance of the ecological context for mycorrhizal functions, as recently comprehensively shown by Hoeksema et al. (2010), was also evident for the impact of the roadside AM fungal community. Treating agricultural soil with roadside inoculum mediated a neutral effect on plant growth compared to the non-mycorrhizal treatment.

Consequently, I found AM fungal assemblages having neutral to negative effects on *A. artemisiifolia* in more fertile agricultural soil, while the function in nutrient limited roadside soil was mutualistic as suggested by the trade balance model (Johnson 2010). I attribute these contrasting effects of mycorrhizal functions in the experiment to abiotic soil conditions, and also the identity of AM fungi establishing the roots. The effect of soil was evident, since total root colonization was of similar magnitude lower in cornfield compared to roadside soil irrespective of inoculum origin, and although I always inoculated soil with 'full soil strength' inocula. Hence, 'parasitic' or 'commensalistic' functions of AM fungi in cornfield soil were not related to a different degree in mycorrhizal root colonization, but likely to different taxa of AM fungi establishing in the roots. Our assessment of fungal structures showed broad shifts in AM fungal taxa colonizing the plant roots depending on soil and inoculum identity, which may have also been influenced by cultivation bias (Sýkorová et al. 2007). Overall, the colonization levels of coarse AM fungi and FE differed significantly among all combinations of soil and mycorrhizal inoculum tested. The different inocula compositions, moreover, were reflected in a trade-off implying that when roots were highly colonized with coarse AM fungi, colonization with FE was low and vice versa. Therefore, divergent functions of AM fungal assemblages in cornfield soil were related to AM fungal taxa, which had colonized the roots. This finding, however, is not surprising given the fact that mycorrhizal function is known to vary among AM fungal genotypes (van der Heijden et al. 1998; Klironomos 2003; Munkvold et al. 2004; Antunes et al. 2011), as well as families (Powell et al. 2009).

Moreover, I found that the negative effect of the mycorrhizal symbiosis in agricultural soil inoculated with the 'local' AM fungal community coincided with increased root fineness compared to inoculation with 'foreign' AM fungi. This demonstrates that the presence of less cooperative AM fungal partners can result in greater branching enabling plants to acquire soil nutrients more directly via fine roots, which is advantageous in nutrient rich soils. For example, Schultz et al. (2001) reported that *Andropogon gerardii* had a more branched and finer root system, and was also less dependent upon mycorrhizal symbiosis, when genotypes came from high-nutrient soil. Similar relationships were also found for plant populations of *Hypericum perforatum*, which had greater root fineness and reduced mycorrhizal responsiveness in the non-native North American range (Seifert et al. 2009). Moreover, *A. artemisiifolia* may be less dependent upon mycorrhizal symbiosis in the new range than expected (Fumanal et al. 2006) because I found no evidence for root systems to be modified towards greater coarseness in the presence of mycorrhizal fungi compared to the control, which often is an indication of mycorrhizal dependency in other plant species (e.g. Hetrick 1991; Berta et al. 2002).

2.5.2 Adaptation to soil environment and plant origin effects

I found no evidence for co-adaptation between *A. artemisiifolia* and AM fungi in the new range because plants both originating from roadside and cornfield seeds grew equally with the respective AM fungal communities. However, plants with roadside origin performed significantly better in roadside soil, which demonstrates adaptation to the abiotic soil environment in the introduced range. In the native range, adaptation of *A. artemisiifolia* to roadside conditions, in particular to high salinity concentrations in roadside soils, has also been reported for the germination behavior of seeds (DiTommaso 2004). Further, results of a few studies indicated adaptation of *A. artemisiifolia* to climate. Recently, Hodgins and Rieseberg (2011) demonstrated adaptation to latitude and climate variables by comparing European with North American populations in common garden experiments. In addition, Chun et al. (2011) found effects of geographic location on reproductive allocation in introduced French populations. In a reciprocal transplant experiment, moreover, weather was found to have a huge effect on re-growth of *A. artemisiifolia* plants threatened with the herbicide imazethapyr, although the reproductive potential of the plants depended on seed origin, suggesting that genetic differences may be a result of evolution of different ecotypes (Leif et al. 2000).

Male function also indicated adaptation of plants with cornfield origin to 'local' soil. In cornfield soil, plants originating from cornfield seeds had significantly more male flower biomass (on average 35 %) than those from roadside seeds. Plant origin aside, more fertile soil conditions strongly increased resource allocation to male inflorescences, thus plants had 2.5 times greater male biomass, and flowered earlier. The correlation of greater plant biomass with a shift towards maleness in nutrient-rich soils was also observed by Ackerly and Jasieński (1990). In a recent study, Friedman and Barrett (2011) showed that sex allocation also varies with light conditions, where plants grown in the sun had higher male flower production.

Plant origin strongly affected root morphology. Plants with roadside origin had significantly finer root systems with smaller root diameters in all treatment situations tested compared to those with cornfield origin. This may result from genetic differences between the populations or maternal effects. Compared to plants in cornfield soil, conspecifics growing in roadside soil produced root systems with smaller root diameters and proportionally greater fine root length: apparently, these root alterations were more efficient in nutrient acquisition in nutrient poor roadside soil because roots explored a larger soil volume per unit root surface area (Gahoonia and Nielsen 2003).

2.5.3 Conclusion

This study shows that performance of non-native *A. artemisiifolia* might be significantly influenced by mycorrhizal functions, leading to a positive effect in less fertile roadside soil and parasitism in more fertile cornfield soil *sensu* Johnson (2010). The effects of natural AM fungal assemblages are strongly soil context dependent (Hoeksema et al. 2010), thus mycorrhizal functions may be unpredictable when AM fungal communities are introduced to novel soils.

For the establishment phase, local adaptation between *A. artemisiifolia* and natural AM fungal assemblages may be of no relevance, highlighting the low host-specificity of AM associations (Richardson et al. 2000). In harsh environmental conditions such as roadside habitats, however, adaptation of *A. artemisiifolia* to soil may play a crucial role in the early stages of invasion, as well as the symbiosis with AM fungi enhancing plant growth and fitness, thus promoting invasive spread. Along the road, therefore, co-adaptation between AM fungi and *A. artemisiifolia* may not be utterly out of the question in later stages of the invasion process (Thompson 2005; Pringle et al. 2009; Hoeksema 2010).

2.6 References

Ackerly DD, Jasieński M (1990) Size-dependent variation of gender in high density stands of the monoecious annual, *Ambrosia artemisiifolia* (Asteraceae). Oecologia 82:474-477.

Allison VJ (2002) Nutrients, arbuscular mycorrhizas and competition interact to influence seed production and germination success in *Achillea millefolium*. Funct Ecol 16:742-749.

Allen MF (1991) The Ecology of Mycorrhizae. Cambridge Studies in Ecology series. Cambridge University Press, Cambridge.

Antunes PM, Koch AM, Morton JB, Rillig MC, Klironomos JN (2011) Evidence for functional divergence in arbuscular mycorrhizal fungi from contrasting climatic origins. New Phytol 189:507-514.

Berta G, Fusconi A, Hooker J (2002) Arbuscular mycorrhizal modifications to plant root systems: scale, mechanisms and consequences. In: Gianinazzi S, Schuepp H, Barea JM, Haselwandter K (eds) Mycorrhizal technology in agriculture, Birkhäuser Verlag, Basel, pp. 71-86.

Bever JD, Richardson SC, Lawrence BM, Holmes J, Watson M (2009) Preferential allocation to beneficial symbiont with spatial structure maintains mycorrhizal mutualism. Ecol Lett 12:13-21.

Blume HP, Deller B, Furtmann K, Leschber R, Paetz A, Wilke BM (2000) In: DIN, Deutsches Institut für Normung e.V. (ed) Handbuch der Bodenuntersuchung. Beuth Verlag, Berlin, pp. HBU 3.4.1.30.2a.

Borowicz VA (2001) Do arbuscular mycorrhizal fungi alter plant-pathogen relations? Ecology 82:3057-3068.

Buswell JM, Moles AT, Hartley S (2011) Is rapid evolution common in introduced plant species? J Ecol 99:214-224.

Brandes D, Nitzsche J (2006) Biology, introduction, dispersal, and distribution of common ragweed (*Ambrosia artemisiifolia* L.) with special regard to Germany. Nachr.bl Dtsch Pflanzenschutzd 58:286-291.

Chun YJ, le Corre V, Bretagnolle F (2011) Adaptive divergence for a fitness-related trait among invasive *Ambrosia artemisiifolia* populations in France. Mol Ecol 20:1378-1388.

Colautti RI, Maron JL, Barrett SCH (2009) Common garden comparisons of native and introduced plant populations: latitudinal clines can obscure evolutionary inferences. Evol Appl 2:187-199.

Crowell HF, Boerner REJ (1988) Influences of mycorrhizae and phosphorus on belowground competition between two old-field annuals. Environ Exper Bot 28:381-392.

Deen W, Hunt T, Swanton CJ (1998) Influence of temperature, photoperiod, and irradiance on the phenological development of common ragweed (*Ambrosia artemisiifolia*). Weed Sci 46:555-560.

DiTommaso A (2004) Germination behavior of common ragweed (*Ambrosia artemisiifolia*) populations across a range of salinities. Weed Sci 52:1002-1009.

Doerr SH (1998) On standardizing the 'water drop penetration time' and the 'molarity of an ethanol droplet' techniques to classify soil hydrophobicity: a case study using medium textured soils. Earth Surf Proc Land 23:663-668.

Friedman J, Barrett SCH (2011) Genetic and environmental control of temporal and size-dependent sex allocation in a wind-pollinated plant. Evolution 65:2061-2074.

Fumanal B, Plenchette C, Chauvel B, Bretagnolle F (2006) Which role can arbuscular mycorrhizal fungi play in the facilitation of *Ambrosia artemisiifolia* L. invasion in France? Mycorrhiza 17:25-35.

Gahoonia TS, Nielsen NE (2003) Root traits as tools for creating phosphorus efficient crop varieties. Plant Soil 260:47-57.

Gotelli NJ, Ellison AM (2004) A Primer of Ecological Statistics. Sinauer Associates, Sunderland, MA.

van der Heijden MGA, Klironomos JN, Ursic M, Moutoglis P, Streitwolf-Engel R, Boller T, Wiemken A, Sanders IR (1998) Mycorrhizal fungal diversity determines plant biodiversity, ecosystem variability and productivity. Nature 396:69-72.

Helgason T, Fitter AH (2009) Natural selection and the evolutionary ecology of the arbuscular mycorrhizal fungi (Phylum Glomeromycota). J Exp Bot 60:2465-2480.

Helgason T, Merryweather JW, Denison J, Wilson P, Young JPW, Fitter AH (2002) Selectivity and functional diversity in arbuscular mycorrhizas of co-occurring fungi and plants from a temperate deciduous woodland. J Ecol 90:371-384.

Hetrick BAD (1991) Mycorrhizas and root architecture. Experientia 47:355-362.

Hoeksema JD (2010) Ongoing coevolution in mycorrhizal interactions. New Phytol 187:286-300.

Hoeksema JD, Chaudhary VB, Gehring, CA, Johnson NC, Karst J, Koide RT, Pringle A, Zabinski C, Bever JD, Moore JC, Wilson GWT, Klironomos JN, Umbanhowar J (2010) A meta-analysis of context-dependency in plant response to inoculation with mycorrhizal fungi. Ecol Lett 13:394-407.

Hodgins KA, Rieseberg L (2011) Genetic differentiation in life-history traits of introduced and native common ragweed (*Ambrosia artemisiifolia*) populations. J Evolution Biol 24:2731-2749.

Ji B, Bentivenga SP, Casper BB (2010) Evidence for ecological matching of whole AM fungal communities to the local plant-soil environment. Ecology 91:3037-3046.

Johnson NC (1993) Can fertilization of soil select less mutualistic mycorrhizae? Ecological Appl 3:749-757.

Johnson NC, Graham JH, Smith FA (1997) Functioning of mycorrhizal associations along the mutualism-parasitism continuum. New Phytol 135:575-585.

Johnson NC (2010) Resource stoichiometry elucidates the structure and function of arbuscular mycorrhizas across scales. New Phytol 185:631-647.

Johnson NC, Wilson GWT, Bowker MA, Wilson JA, Miller MR (2010) Resource limitation is a driver of local adaptation in mycorrhizal symbioses. PNAS 107:2093-2098.

Joly M, Bertrand P, Gbangou RY, White MC, Dubé J, Lavoie C (2011) Paving the way for invasive species: road type and the spread of common ragweed (*Ambrosia artemisiifolia*). Environ Manage 148:514-522.

Kawecki TJ, Ebert D (2004) Conceptual issues in local adaptation. Ecol Lett 7:1225-1241.

Kiers ET, Duhamel M, Beesetty Y, Mensah JA, Franken O, Verbruggen E, Fellbaum CR, Kowalchuk GA, Hart MM, Bago A, Palmer TM, West SA, Vandenkoornhuyse P, Jansa J, Bücking H (2011) Reciprocal rewards stabilize cooperation in the mycorrhizal symbiosis. Science 333:880-882.

Klironomos JN, Hart MM (2002) Colonization of roots by arbuscular mycorrhizal fungi using different sources of inoculum. Mycorrhiza 12:181-184.

Klironomos JN (2003) Variation in plant response to native and exotic arbuscular mycorrhizal fungi. Ecology 84:2292-2301.

Klironomos JN, Moutoglis P, Kendrick B, Widden P (1993) A comparison of spatial heterogeneity of vesicular-arbuscular mycorrhizal fungi in two maple-forest soils. Can J Botany 71:1472-1480.

Koide RT, Dickie IA (2002) Effects of mycorrhizal fungi on plant populations. Plant Soil 244:307-317.

Koide RT, Li M (1989) Appropriate controls for vesicular-arbuscular mycorrhizal research. New Phytol 111:35-44.

Legendre P, Legendre L (1998) Numerical Ecology, 2nd English edn. Elsevier Science BV, Amsterdam.

Leif JW, Vollmer JL, Hartberg TJ, Ballard TO (2000) Growth and response of common ragweed (*Ambrosia artemisiifolia*) ecotypes to imazethapyr. Weed Technol 14:150-155.

van Lierop WM, MacKenzie AF (1977) Soil pH measurement and its application to organic soils. Can J Soil Sci 57:55-64.

Lu X, Koide RT (1994) The effects of mycorrhizal infection on components of plant growth and reproduction. New Phytol 128:211-218.

Marler MJ, Zabinski CA, Callaway RM (1999) Mycorrhizae indirectly enhance competitive effects of an invasive forb on a native bunchgrass. Ecology 80:1180-1186.

McGonigle TP, Miller MH, Evans DG, Fairchild GL, Swan JA (1990) A new method which gives an objective measure of colonization of roots by vesicular arbuscular mycorrhizal fungi. New Phytol 115:495-501.

Mummey DL, Rillig MC (2006) The invasive plant species *Centaurea maculosa* alters arbuscular mycorrhizal fungal communities in the field. Plant Soil 288:81-90.

Munkvold L, Kjøller R, Vestberg M, Rosendahl S, Jakobsen I (2004) High functional diversity within species of arbuscular mycorrhizal fungi. New Phytol 164:357-364.

Newsham KK, Fitter AH, Watkinson AR (1995) Multi-functionality and biodiversity in arbuscular mycorrhizas. Trends Ecol Evol 10:407-411.

Pánková H, Münzbergová Z, Rydlová J, Vosátka M (2008) Differences in AM fungal root colonization between populations of perennial *Aster* species have genetic reasons. Oecologia 157:211-220.

Pearson JN, Jakobsen I (1993) Symbiotic exchange of carbon and phosphorus between cucumber and three arbuscular mycorrhizal fungi. New Phytol 124:481-488.

Powell JR, Parrent JL, Hart MM, Klironomos JN, Rillig MC, Maherali H (2009) Phylogenetic trait conservatism and the evolution of functional trade-offs in arbuscular mycorrhizal fungi. P R Soc B 276:4237-4245.

Pringle A, Bever JD, Gardes M, Parrent JL, Rillig MC, Klironomos JN (2009) Mycorrhizal symbioses and plant invasions. Annu Rev Ecol Evol S 40:699-715.

R Development Core Team (2009). R: A language and environment for statistical computing. R Foundation for Statistical Computing, Vienna, Austria. ISBN 3-900051-07-0, URL http://www.R-project.org.

Richardson DM, Allsopp N, D'Antonio CM, Milton SJ, Rejmánek M (2000) Plant invasions – the role of mutualisms. Biol Rev Camb Philos 75:65-93.

Sakai AK, Allendorf FW, Holt JS, Lodge DM, Molofsky J, With KA, Baughman S, Cabin RJ, Cohen JE, Ellstrand NC, McCauley DE, O'Neil P, Parker IM, Thompson JN, Weller SG (2001) The population biology of invasive species. Annu Rev Ecol Syst 32:305-332.

Schultz PA, Miller RM, Jastrow JD, Rivetta CV, Bever JD (2001) Evidence of a mycorrhizal mechanism for the adaptation of *Andropogon gerardii* (Poaceae) to high- and low-nutrient prairies. Am J Bot 88:1650-1656.

Seifert EK, Bever JD, Maron, JL (2009) Evidence for the evolution of reduced mycorrhizal dependence during plant invasion. Ecology 90:1055-1062.

Smith SE, Read DJ (2008) Mycorrhizal Symbiosis, 3rd edn. Academic Press, London, UK.

Sun Y, Li XL, Feng G (2008) Effect of arbuscular mycorrhizal colonization on ecological functional traits of ephemerals in the Gurbantonggut desert. Symbiosis 46:121-127.

Sýkorová Z, Ineichen K, Wiemken A, Redecker D (2007) The cultivation bias: different communities of arbuscular mycorrhizal fungi detected in roots from the field, from bait plants transplanted to the field, and from a greenhouse trap experiment. Mycorrhiza 18:1-14.

Thippayarugs S, Bansal M, Abbott LK (1999) Morphology and infectivity of fine endophyte in a mediterranean environment. Mycol Res 103:1369-1379.

Thompson JN (2005) The Geographic Mosaic of Coevolution. University of Chicago Press, Chicago, USA.

Thompson JN (1999) The evolution of species interactions. Science 284:2116-2118.

Verbruggen E, Kiers, ET (2010) Evolutionary ecology of mycorrhizal functional diversity in agricultural systems. Evol Appl 3:547-560.

Veresoglou SD, Rillig MC (2012) Suppression of fungal and nematode plant pathogens through arbuscular mycorrhizal fungi. Biology Lett 8:214-216.

Vierheilig H, Coughlan AP, Wyss U, Piche Y (1998) Ink and vinegar, a simple staining technique for arbuscular-mycorrhizal fungi. Appl Environ Microb 12:5004-5007.

Vogelsang KM, Bever JD (2009) Mycorrhizal densities decline in association with nonnative plants and contribute to plant invasion. Ecology 90:399-407.

Zhang Q, Yang R, Tang J, Yang H, Hu S, Chen X (2010) Positive feedback between mycorrhizal fungi and plants influences plant invasion success and resistance to invasion. PloS ONE 5:e12380. doi:10.1371/journal.pone.0012380.

2.7 Tables and figures

Table I.1 Characteristics of the soils used in the experiment. Analyses of soil pH (as measured by deionized water and exchangeable in a 1:3 soil:0.01 M $CaCl_2$ suspension), water repellency, extractable contents of nitrogen (N), carbon (C), and plant available phosphorus (P) refer to soil properties after steaming. Values represent means ± SE (n = 3, except for water repellency n = 5). *P*-values relate to *t*-tests for independent samples. Values in bold indicate significance at $P < 0.05$.

Soil	Roadside	Cornfield	*P*
pH (deionized H_2O)	5.54 ± 0.06	6.12 ± 0.03	**<0.001**
pH ($CaCl_2$)	4.51 ± 0.03	5.31 ± 0.02	**<0.001**
Water repellency (sec)	12.4 ± 4.2	2.2 ± 1.2	**<0.001**
Total C in %	0.560 ± 0.153	0.581 ± 0.043	0.827
Total N in %	0.028 ± 0.007	0.047 ± 0.003	**0.013**
Plant available P (mg per 100 g soil)	6.25 ± 2.50	8.33 ± 2.20	0.339

Table I.2 Eigenvectors of the first five principal components (cumulative % variance explained > 97) of the Principal Component Analyses (PCAs) performed on the correlation matrix of three different sets of plant traits listed in the first column.

PCA Plant biomass	PC1	PC2	PC3	PC4	PC5
Shoot biomass	0.475	-0.165	0.178	-0.054	0.365
Root biomass	0.434	-0.333	0.140	-0.031	0.529
Reproduction 5 weeks	0.220	0.634	-0.125	0.687	0.245
Total seed number	0.451	0.179	-0.294	-0.248	-0.297
Ripe seed wt.	0.166	0.571	0.629	-0.442	-0.093
Immat. seed wt.	0.415	0.056	-0.553	-0.233	-0.186
Male flower wt.	0.367	-0.315	0.382	0.461	-0.629
Proportion of total SS	53%	20%	12%	7%	5%

Table I.2 continued.

PCA Root traits	PC1	PC2	PC3	PC4	PC5
Root length per volume	0.443	0.059	-0.310	-0.221	0.017
Root surface area	0.443	-0.137	0.061	-0.184	0.426
Fine root volume	0.288	0.584	0.019	0.139	-0.661
Fine root length	0.264	0.613	0.114	0.293	0.570
Coarse root volume	0.384	-0.212	0.816	-0.182	-0.176
Coarse root length	0.428	-0.135	-0.451	-0.357	-0.132
Root diameter	0.351	-0.445	-0.131	0.808	-0.089
Proportion of total SS	70%	25%	4%	0.01%	0.001%
PCA Mycorrhization					
Total root colonization	-0.511	0.112	-0.170	0.367	-0.208
coarse AMF hyphae	-0.317	0.451	-0.233	-0.046	-0.611
coarse AMF arbuscules	-0.264	0.476	-0.415	-0.250	0.679
coarse AMF vesicles	-0.249	0.431	0.857	0.016	0.128
FE hyphae	-0.421	-0.348	0.022	0.200	0.062
FE arbuscules	-0.408	-0.362	0.021	0.325	0.289
FE vesicles	-0.405	-0.345	0.095	-0.809	-0.136
Proportion of total SS	51%	42%	5%	1%	0.01%

SS, sum of squares; Ripe seed wt., weight of ripe seeds; Immat. seed wt., weight of immature seeds; Male flower wt., weight of male inflorescences; FE, fine endophytes (a special group of AM fungi); coarse AMF, 'normal' AM fungi (termed here as 'coarse' AM fungi).

Table I.3 Analyses of variance (ANOVAs) on the first principal component score (PC1) of the Principal Component Analyses (PCAs) on plant biomass, root traits and mycorrhization of *Ambrosia artemisiifolia* in the experiment, with P.ori (plant origin), Soil, and Myc. (mycorrhizal treatment) as factors. Values in bold indicate significance at $P < 0.05$.

PC1 of PCA on		Plant Biomass		Root traits		Mycorrhization	
Factors	df	F	P	F	P	F	P
P.ori	1	1.98	0.162	1.93	0.170	0.54	0.465
Soil	1	**942.65**	**<0.001**	**92.18**	**<0.001**	**65.70**	**<0.001**
Myc.	2	0.72	0.488	2.22	0.118	**318.00**	**<0.001**
P.ori x Soil	1	**5.33**	**0.022**	**4.41**	**0.040**	0.74	0.392
P.ori x Myc.	2	0.24	0.786	1.44	0.246	0.22	0.803
Soil x Myc.	2	**17.00**	**<0.001**	**3.75**	**0.030**	**16.34**	**<0.001**
P.ori x Soil x Myc.	2	1.10	0.337	0.11	0.897	0.22	0.804
Residuals			132		54		132

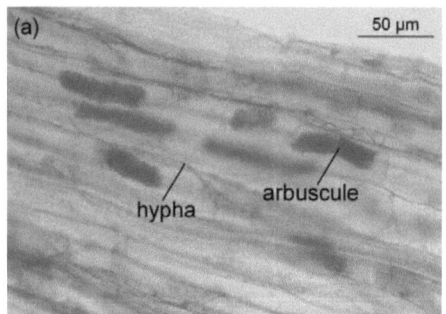
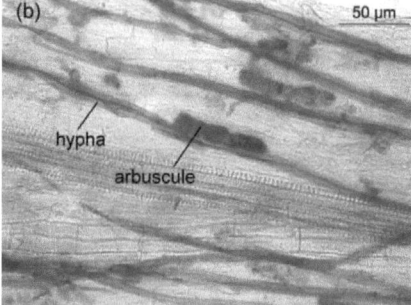

Figure I.1 Colonization with arbuscular mycorrhizal (AM) fungi in roots of *Ambrosia artemisiifolia* in the experiment: (a) colonization with fine endophytes (a special group of AM fungi); (b) colonization with the 'normal' AM fungi (here termed as 'coarse' AM fungi). Roots were stained with ink-vinegar method (Vierheilig et al. 1998). Photograph: Cornelia Bäucker.

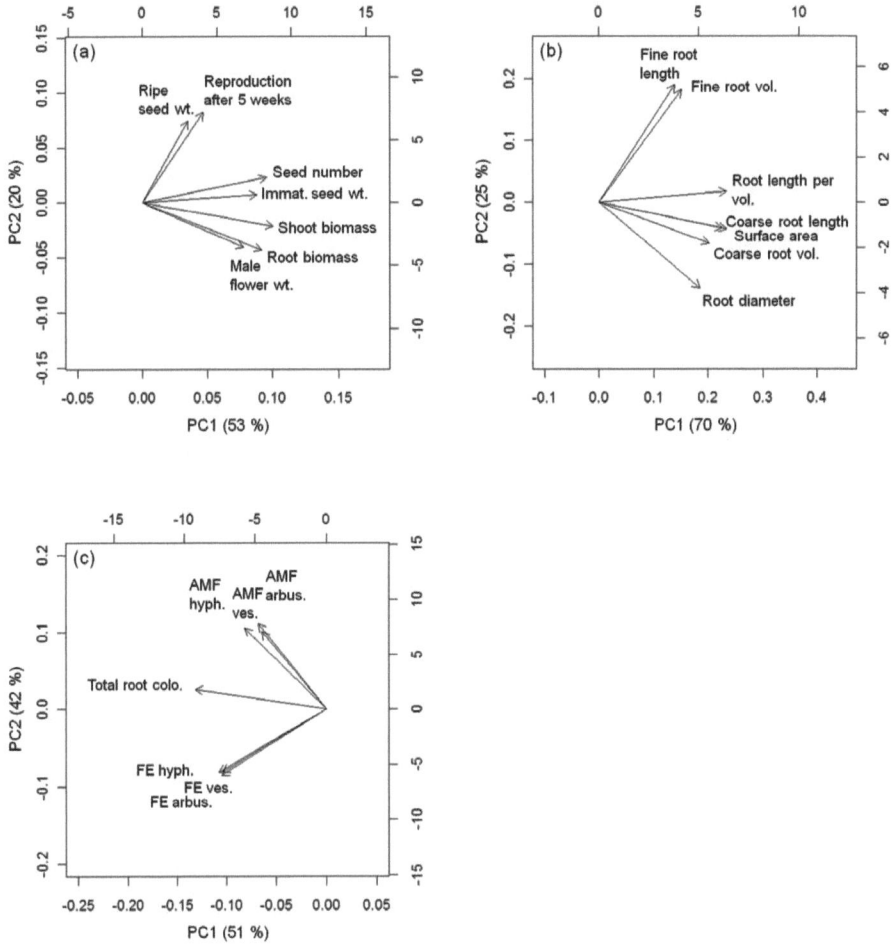

Figure I.2 Loading plots of the Principal Component Analyses (PCAs) performed on the correlation matrix of variables of different plant trait aspects including always seven response variables as listed in Table 1.2: (a) biomass traits; (b) root traits; (c) mycorrhization. The variable arrow coordinates are built from PC1 and PC2 eigenvector coefficients (Table I.3) and visualize how variables correlate with each other.

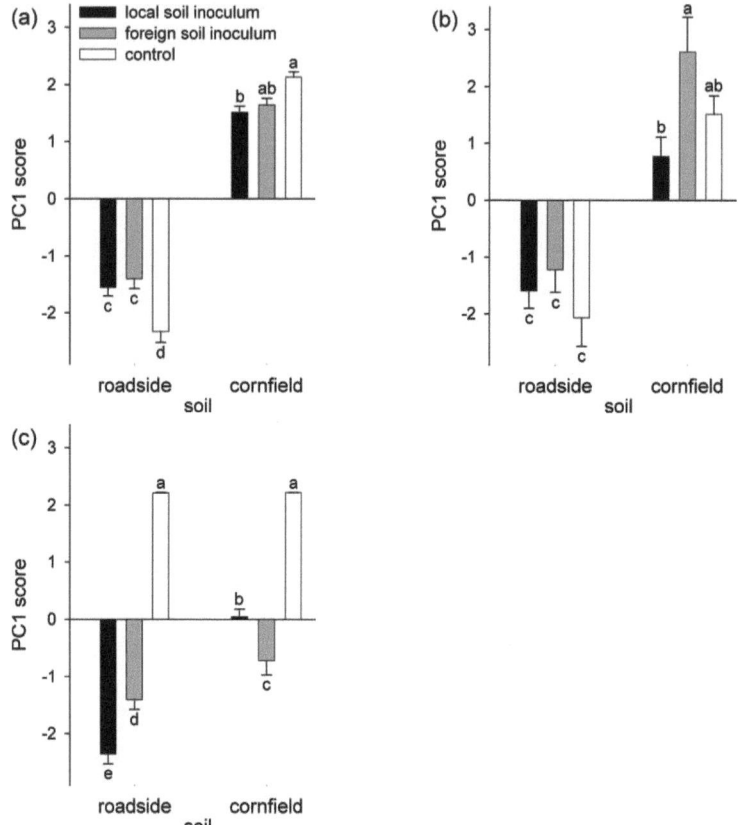

Figure I.3 Responses to soil and mycorrhizal treatments of *Ambrosia artemisiifolia* in the experiment: PC1 scores refer to the PCA on plant biomass traits (a); root traits (b); mycorrhization (c). Bar plots of mean (± SE) indicate PC1 score variation in response to treatments of soils inoculated with 'local' soil AM fungal inoculum (black), 'foreign' soil AM fungal inoculum (grey), and non-mycorrhizal controls (white). For (a) and (b) higher or less negative PC1 scores correspond to greater biomass and bigger root systems of greater coarseness, respectively. In (c) lower PC1 scores mean higher root colonization both with AM fungal hyphae in total and structures of fine endophytes (hyphae, arbuscules, vesicles). Different lower case letters on the bars indicate a significant difference ($P < 0.05$) among treatment groups according to Tukey's HSD test.

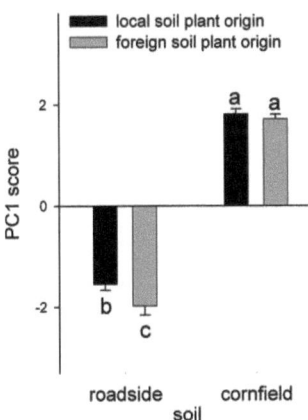

Figure I.4 First principal component score of the PCA on biomass traits of *Ambrosia artemisiifolia* indicating local adaptation to roadside soil in the experiment: higher PC1 scores (means ± SE) indicate greater plant biomass. Black bars correspond to the combination of soil with the respective 'local' plant origin and grey bars to 'foreign' origin. 'Local soil plant origin' means that plants grew in their respective own soil, i.e. plants from the roadside in roadside soil and plants from the cornfield in cornfield soil, respectively. The term 'foreign plant soil origin' refers to new combinations of plants and soil, i.e. when plants from the cornfield grew in roadside soil and plants from the roadside grew in cornfield soil, respectively. Different lower case letters on the bars indicate a significant difference ($P < 0.05$) among treatment groups according to Tukey's HSD test.

3 STUDY II: NON-NATIVE *AMBROSIA ARTEMISIIFOLIA* ARE MORE INFLUENCED BY RELATIVE DENSITY AND IDENTITY OF NEIGHBORING PLANT SPECIES THAN ARBUSCULAR MYCORRHIZA

3.1 Abstract

Arbuscular mycorrhizal (AM) fungi can play a crucial role in plant invasions by mediating a competitive advantage of invasive over native species. Since invasive *Ambrosia artemisiifolia* has been proposed to be facilitated by AM fungi, this study investigated whether its competitive ability is enhanced by the symbiosis with AM fungi under a maintained ecological context of soil and AM fungal communities. I studied *A. artemisiifolia* grown together with one of four co-existing mycorrhizal plant species in a 1:4 (target) or 4:1 (challenger) relative density. The neighbor species were *Conyza canadensis*, *Daucus carota*, *Artemisia vulgaris*, and *Tanacetum vulgare*. *D. carota* and *A. artemisiifolia* were further tested in pairwise competition. *A. artemisiifolia* was always dominant and I found no indication that AM fungi influenced the interspecific competitive outcome under low soil fertility conditions. However, AM fungi had a significant amplifying effect on intraspecific competition of *A. artemisiifolia*. My findings suggest that the competitive ability of *A. artemisiifolia* is very weakly influenced by AM fungi in the presence of other mycorrhizal plant competitors under nutrient poor soil conditions in Central Europe. The invasive success of *A. artemisiifolia* may be related to mechanisms other than facilitation by AM fungi.

3.2 Introduction

Invasive plant species are 'naturalized plants that produce reproductive offspring, often in very large numbers, at considerable distances from parent plants and thus have the potential to spread over a considerable area' (Richardson et al. 2000a). Worldwide, numerous ecosystems are affected by invasive plants in their functioning in a multitude of ways (Wardle et al. 2011). Invasive species, moreover, can damage ecosystem services that are fundamental to human well-being resulting in substantial economic costs (e.g. Pimentel et al. 2005; Pejchar and Mooney 2009; Vilà et al. 2010). Therefore, it is of increasing urgency to better understand the mechanisms involved in the invasion process as human activities, such as international trade, transport and travel, which cause species dispersal into new ranges, continue to expand (Keller et al. 2011).

Mechanisms by which invasive plants successfully spread into new areas may primarily be related to interactions of the invasive species with their new environment (Jeschke et al. 2012). Therefore, aboveground mutualistic interactions, such as pollination or seed dispersal, may play an important role for the success of introduced plants (Richardson et al. 2000b). In addition, belowground symbioses, such as mycorrhizal relationships, have also been shown to be a critical aspect (Richardson et al. 2000b; Pringle et al. 2009; Shah et al. 2009). For example, arbuscular mycorrhizal (AM) fungi, which form complex underground hyphal networks with roots of around two-thirds of plant species (Helgason and Fitter 2009), and provide multiple functions, especially increased phosphorus uptake to the plant in exchange for photoassimilates (Newsham et al. 1995), are known to be a crucial factor in some plant invasions. One way by which invasive plants can interact with natural AM fungal communities is by disrupting the formation of mycorrhizal associations upon which many native plant species depend (Stinson et al. 2006; Meinhardt and Gehring 2012). Another important possibility how invasive plants can interact with AM fungi has been described as facilitation (Reinhart and Callaway 2006; Shah et al. 2009), whereby invasive plants are positively influenced by the AM fungal association of the new range to the detriment of native species.

Several studies have documented that AM fungi can contribute to the dominance of invasive over native plants by altering competitive interactions (Marler et al. 1999; Callaway et al. 2003; Callaway et al. 2004; Callaway et al. 2005; Shah et al. 2008; Wilson et al. 2012; but see Bray et al. 2003; Emery and Rudgers 2012). However, the mechanism underlying this increased competitive ability of invasive plants has not been fully elucidated; it might be caused by changes in AM fungal community composition in the presence of the invasive species (Mummey and Rillig 2006; Hawkes et al. 2006; Zhang et al. 2007), or increased P-uptake of the invasive plant via AM fungal mycelia (Zabinski et al. 2002). Further, the outcome of competitive interactions between plants has been shown to be influenced by identity of AM fungi (Scheublin et al. 2007; Shah et al. 2008). Therefore, the ecological context of studies investigating the effect of AM fungi on competitive interactions between invasive and native plant species has to be carefully considered, as recently demonstrated by Hoeksema et al. (2010).

Here, I study the competitive ability of *Ambrosia artemisiifolia* L. (Astereceae), an annual plant native to North America. There, the species is known to be dominant during early stages of old-field succession in many parts of the eastern and midwestern United States (Bazzaz and Mezga 1973). Introduced to other continents, the species successfully spreads, and poses a risk for human health because it can cause hay fever through its pollen (e.g. Brandes and Nitzsche 2006). A special trait of the monoecious genus Ambrosia is that sexes are located in different types of flowers on the individuals; hence, *A. artemisiifolia* allows quantification of resource allocation to functional genders (Friedman and Barrett 2011). The plant occurs in different types of disturbed habitats, for example, roadsides, construction sites, waste lands, and agricultural fields, where it is an important weed (Brandes and Nitzsche 2006). Moreover, *A. artemisiifolia* is a mycorrhizal plant and its spread has been suggested to be facilitated by the symbiosis with arbuscular mycorrhizal fungi in Europe (Fumanal et al. 2006). In a greenhouse experiment, Crowell and Boerner (1988) showed that *A. artemisiifolia* responds positively to inoculation with mycorrhizal fungi: plant shoot and total biomass were more than 20 times increased in the presence of AM fungi (the AM fungal taxon tested was *Glomus etunicatum*). The same study, further, reports that interspecific competition with the non-mycorrhizal *Brassica nigra* was stronger than intraspecific competition in the presence of AM fungi: here, AM fungal inoculation conferred no advantage to the mycorrhizal competitor, which is rather atypical for competitive situations of mycorrhizal with non-mycorrhizal plants (Moora and Zobel 2010). Other studies on competition, however, found that *A. artemisiifolia* had a large suppressive effect on other co-existing species; hence, *A. artemisiifolia* was competitively dominant (Miller and Werner 1987; Miller 1994; but see Leskovšek et al. 2012).

This study aimed to investigate whether the competitive ability of *A. artemisiifolia* is enhanced by the symbiosis with natural AM fungi in the presence of co-occurring plant species in the new European range under conditions of a maintained ecological context of soil and AM fungal communities. The hypotheses were: 1) AM fungi should confer a competitive advantage on *A. artemisiifolia* in the presence of other mycorrhizal competitors; 2) the growth of neighbors should be more decreased in the presence of a high than a low density of *A. artemisiifolia;* 3) the competitive outcome should be independent of the neighboring plant species.

3.3 Materials and methods

I conducted two greenhouse experiments with comparatively high realism, i.e., I always maintained the ecological context of soil and natural AM fungal communities (Hoeksema et al. 2010). The first experiment focused on *A. artemisiifolia* in target and challenger arrangements in combination with four different neighboring plant species in the presence or absence of AM fungi: hereafter termed target–challenger experiment. As neighbor species I chose *Conyza canadensis* L. (Asteraceae), *Daucus carota* L. (Apiaceae), *Artemisia vulgaris* L. (Asteraceae), and *Tanacetum vulgare* L. (Asteraceae): all co-existing with *A. artemisiifolia* and mycorrhizal. Because this target–challenger experiment indicated that the mycorrhizal association tended to have a growth reducing effect on *A. artemisiifolia* in the presence of *D. carota,* I conducted a pairwise competition experiment with these two plant species: hereafter referred to as pairwise competition experiment.

3.3.1 Target–challenger experiment

The experiment had a fully 2 x 4 x 2 factorial design and was replicated seven times. It was comprised of two relative density arrangements of *A. artemisiifolia* (target and challenger), four neighboring plant species (*C. canadensis*, *D. carota*, *A. vulgaris*, *T. vulgare*), and two soil treatments (natural AM fungal community and non-mycorrhizal control). All arrangements were interspecific and had five plants per pot: one target plant in the center surrounded by four challenger plants belonging to the same plant species (Figure II.1). Thus, the two different density levels of *A. artemisiifolia* were realized by either having *A. artemisiifolia* as target in a ratio of 1:4 (Figure II.1a) or as challenger in a ratio of 4:1 (Figure II.1b) in comparison to each of four selected neighboring plant species of the new range. All neighboring plant species tested are mycorrhizal plants, but have some differences in life span and growth form: *C. canadensis* is annual and forms at first a leaf rosette in spring. *D. carota* is an upright biennial, and *A. vulgaris* and *T. vulgare* are upright perennials. I found these plant species co-occurring with the annual, upright growing *A. artemisiifolia*, which is in accordance with other observations for *A. artemisiifolia* in Middle Europe (Brandes and Nitzsche 2007).

In October 2007, I collected seeds of *A. artemisiifolia* and the four co-occurring plant species from plants from a ruderal site in Berlin, Germany (52°28'30.00"N, 13°21'46.45"E). The seeds were stored at room temperature or minus 20 °C (*A. artemisiifolia*). In February 2008, seeds of all plants species were pre-germinated in sterile playground sand in a greenhouse (day/night temperature 22/18 °C; ambient light conditions). Within one week all plant species started to germinate. Two weeks later I used the seedlings in the experiment.

The soil used in the experiment came from a field research station of the Humboldt Universität Berlin close to Berlin: Thyrow (52°15' N, 13°14' E). The Thyrow soil is characterized as sandy to silty sand soil (Ellmer et al. 2000), and has a low fertility (organic carbon = 0.52 %; total nitrogen = 0.04 %), as well as an acidic pH (soil pH = 5.2). I used that sandy soil, because competition has been shown to be greater for soil nutrients than for light under such nutrient poor soil conditions (Rebele 2000); hence, by using a low-fertility soil I expected an increased importance of AM fungi. In March 2008, I randomly took soil samples from the top 15 cm of soil. These soil samples were pooled and sieved through a 4 mm sieve. Except for 20 L of soil, which were used for mycorrhizal inoculum production, soil was pasteurized by steaming (Sterilo 7K, Harter Elektotechnik, Schenkenzell, Germany) at 90 °C overnight. For mycorrhizal inoculum, non-steamed soil was thoroughly mixed with water and the supernatant was wet sieved through 500, 212, 53, 38, 20 µm sieves. Filtrate prepared from 53–38 µm soil fraction was used as mycorrhizal inoculum, and filtrate that passed through the 20 µm sieve was collected as microbial wash.

To set up the experiment, seedlings were planted in target and challenger arrangements in 3 L plastic pots filled with 3.5 kg of soil. To ensure that distances between plants were identical, I marked planting positions with a circle template in all 112 pots. After planting, pots intended for mycorrhizal treatment received 50 ml of the extracted mycorrhizal inoculum filtrate (inocula soil:experiment soil ratio 1:10), while an equivalent amount of water was added to the non-mycorrhizal controls. To correct for differences in microbial background communities, all pots received 50 ml of the extracted microbial wash filtrate. Afterwards, pots were covered with a final 125 ml of steamed soil.

Plants grew in a fully randomized arrangement for seven weeks in a greenhouse (photoperiod 16 h; day/night temperature 22/18 °C), and were watered as needed with the same amount of water. At harvest, I found pots heavily penetrated by roots in all arrangements. Because it was impossible to separate roots from each other, I focused on shoot biomass of plant species, as well as male inflorescences and female flowers produced by *A. artemisiifolia*. Shoots were cut away from roots and number of male/female flowers was counted. Shoot weight was determined after drying for five days at 40 °C.

3.3.2 Pairwise competition experiment

Performance of *A. artemisiifolia* and *D. carota* was studied in situations of intra- and interspecific competition in a fully 3 x 2 factorial design. The experiment consisted of three competitive arrangements (intraspecific competition of *A. artemisiifolia*, hereafter AA; intraspecific competition of *D. carota*, hereafter DD; interspecific competition between *A. artemisiifolia* and *D. carota*, hereafter AD) and two AM fungal community treatments (presence, absence). For each treatment combination I set up 12 replicates.

In the experiment, the ecological context was strictly maintained (Hoeksema et al. 2010): plant seeds, soil and AM fungal community came from a same roadside site in Lower Lusatia, Germany (51°44'02.00"N, 14°27'27.10"E). The soil of the habitat was defined as sandy and low fertile (total carbon = 0.56 ± 0.15 %; total nitrogen = 0.03 ± 0.01 %; soil pH = 5.5 ± 0.1) (Bäucker, present book, Study I). In October 2008, I collected seeds from 12 randomly chosen plants. In April 2009, ten soil samples (ca. 10 L each) were taken from the top 12 cm of soil. Soil samples were mixed, sieved (4 mm), and pasteurized by steaming for four hours at 90 °C (Sterilo 1K, Harter Elektotechnik, Schenkenzell, Germany). A portion of soil (15 L) was kept non-steamed and used for production of mycorrhizal inoculum and microbial wash filtrate (wet sieving method as described above). After extraction of AM fungal spores, the mycorrhizal inoculum was cleaned by using the sucrose centrifugation method (Johnson et al. 1999).

The experiment was set up in a greenhouse (photoperiod 16 h; day/night temperature 25/18 °C). To this, plastic pots (1 L) were filled with 700 ml of steamed soil. Corresponding to the three competitive situations (AA, DD, AD), I planted two seedlings (two-week-old; germinated in sterile playground sand) per pot. Planting positions in the pots were marked with a circle template so that distances between seedlings were identical in all experimental units. After planting, 10 ml of mycorrhizal inoculum was added to roots in pots intended for mycorrhizal treatment (non-mycorrhizal controls received 10 ml water). Afterwards, all pots received 60 ml of the extracted microbial wash filtrate, and a final 110 ml of steamed soil. Plants grew in a fully randomized arrangement for six weeks, and were watered as needed (every day or every other day). At harvest, I removed shoot biomass, and carefully washed the root ball and separated roots of the two plants from each other. I determined shoot and root biomass per individuum after drying plant material for six days at 40 °C.

3.3.3 Mycorrhizal colonization in the roots

From both experiments I stained a representative root subsample from the total root biomass. As staining method I used the ink-vinegar procedure (Vierheilig et al. 1998). For this, roots were cut in pieces (1–2 cm) and treated in 10 % KOH for 35 min (water bath 90 °C). After decantation of KOH, roots were thoroughly rinsed with demineralized water, and cooked for another 15 min in ink-vinegar solution (1:1:8 Schaeffer ink, 10 % acetic acid and water). Afterwards, roots were de-stained in demineralized water and stored in lactoglycerol (1:1:1 lactic acid, glycerol and water). Presence of AM hyphae, arbuscules, and vesicles, as well as other root fungal structures were assessed by using the grid-line intersect method (McGonigle et al. 1990) based on 100 intersections per root sample examined at 200X (compound microscope, Leica Microsystems CMS GmbH, Wetzlar, Germany).

3.3.4 Statistical analyses

For the target–challenger experiment I performed standard three-way ANOVAs to test for significance of main effects (relative density arrangement of *A. artemisiifolia*, neighboring plant species, soil treatment) and their interactions. Data were transformed as necessary to meet assumptions of normality and homogeneity of variances, as indicated by Shapiro-Wilk test (on residuals) and Bartlett test, respectively. I used log-transformation for data of vegetative shoot biomass of *A. artemisiifolia* and BoxCox-transformation (exponent 0.4242424) for biomass data of neighboring plant species. Data of number of male inflorescences and female flowers of *A. artemisiifolia* were square-root transformed. To make comparisons between plant species when grown as target (one plant in the center per pot) and challenger (four plants belonging to the same species in a circle per pot), I used the average value of the four challenger plants per pot.

For the pairwise competition experiment of *A. artemisiifolia* and *D. carota* I performed two-way ANOVAs testing for significance of the main effects (competitive situation and soil treatment) and their interaction. The analyses on biomass and mycorrhization of the plant species were computed with two levels for competitive situation (*A. artemisiifolia* with level AA, AD; *D. carota* with level DD, AD). I based the analyses on the mean of the two plants in monoculture and the measure of the respective plant species in the mixed situation. To meet assumptions of normality (Shapiro-Wilk test) and homogeneity of variances (Bartlett test), data of shoot and root biomass of *A. artemisiifolia* and *D. carota* were log-transformed. To test for differences in plant performance depending on presence/absence of mycorrhiza in single situations of intra- and interspecific competition of *A. artemisiifolia* and *D. carota*, I used paired *t*-test's for independent samples. Again, data were analyzed after log-transformation, except for shoot and root biomass of *A. artemisiifolia* in mixture.

Differences between treatment groups were always compared with Tukey HSD post-hoc comparison tests ($P < 0.05$). Computations were performed using R version 2.10.1 (R Development Core Team 2009).

3.4 Results

3.4.1 Target–challenger experiment

I found that the target or challenger arrangement of *A. artemisiifolia* had a substantial influence on all response variables measured (main effect of relative density always $P < 0.001$; Table II.1). When neighboring plant species were planted in the center surrounded by *A. artemisiifolia*, their pooled shoot biomass was on average smaller (mean ± SE; 0.185 ± 0.017 g) than if they grew in a circle (pooled shoot biomass of the mean of all neighbors per pot: 0.524 ± 0.040 g). In contrast, *A. artemisiifolia* produced on average considerably more vegetative shoot biomass per plant as target (3.184 ± 0.181 g) compared to challenger (1.158 ± 0.021 g) (Figure II.3a); hence, both *A. artemisiifolia* and neighboring plant species had on average greater shoot biomass per plant in target-situations of *A. artemisiifolia*. Further, the number of male inflorescences produced by *A. artemisiifolia* was on average almost tripled when it grew as target (7.8 ± 0.9) compared to challenger (2.7 ± 0.2) (Figure II.3b). Conversely, the number of female flowers produced by *A. artemisiifolia* per plant was on average lower in target (0.3 ± 0.1) compared to challenger arrangements (1.7 ± 0.2).

Furthermore, identity of the neighboring plant species also strongly influenced shoot biomass produced by both *A. artemisiifolia* and the four neighboring plant species (significant main effect of neighboring plant species; Table II.1). Moreover, biomass of *A. artemisiifolia* and neighboring plant species depended on if neighboring plants where tested in target or challenger arrangements of *A. artemisiifolia* (significant two-way interaction term between relative density and neighboring plant species; Table II.1). Effects described hereafter are significant, unless otherwise stated. I found that *C. canadensis* performed poorly in all arrangements compared to the other neighboring plant species tested (Figure II.2). Moreover, when *A. artemisiifolia* was grown as challenger, *D. carota*, *A. vulgaris*, and *T. vulgare* showed no significant differences in aboveground biomass. In target arrangements of *A. artemisiifolia*, however, *A. vulgaris* had greater shoot biomass than *D. carota* (arrangement target *A. vulgaris–D. carota*: $P = 0.009$; after Tukey's HSD pairwise comparison test) (Figure II.2). Overall, therefore, *C. canadensis* produced the smallest shoot biomass compared to all other neighboring species, and *D. carota* had less biomass compared *A. vulgaris* and *T. vulgare,* respectively (Figure 2; Supplemental Table A.II.2). Regarding *A. artemisiifolia*, I found that its shoot growth as target and overall was greater when grown together with *C. canadensis* compared to all other combinations of target/challenger arrangements and neighboring plant species tested ($P < 0.05$; Tukey HSD) (for data see Supplemental Tables A.II.2 and A.II.3; Figure II.3a).

Further, a significant three-way interaction term indicated that responses of shoot biomass and maleness of *A. artemisiifolia* when grown as target or challenger were influenced by neighboring plant species and mycorrhizal treatment (Table II.1). In the presence of AM fungi, I found that shoot biomass of target *A. artemisiifolia* tended to be greater only when grown in combination to *C. canadensis* and *T. vulgare*, respectively (Figure II.3a). Conversely, when *A. artemisiifolia* was target to *D. carota*, shoot biomass tended to be decreased in the presence of AM fungi compared to non-mycorrhizal control (Figure II.3a). For combination with *A. vulgaris* as the neighboring plant, I found that shoot growth of target *A. artemisiifolia* was uninfluenced by presence/absence of AM fungi. Furthermore, the different AM fungal effects on *A. artemisiifolia* as a function of competing neighbor species were also found overall, as indicated by a significant interaction term between neighboring plant species and soil treatment: again, *A. artemisiifolia* tended to either increase in shoot biomass (with *C. canadensis* or *T. vulgare*,), decrease (with *D. carota*) or was unaffected (with *A. vulgaris*) in the presence of mycorrhiza (Table II.1; Supplemental Table A.II.4). But, all shoot biomass effects mediated by presence/absence of AM fungi could not be shown with pairwise comparison tests (Tukey HSD; $P > 0.05$).

Considering number of male inflorescences of *A. artemisiifolia*, however, I found that more male flowers were produced when *A. artemisiifolia* grew as target surrounded by *C. canadensis* in the presence of AM fungi compared to absence (arrangement target to *C. canadensis* mycorrhizal–non-mycorrhizal: $P = 0.048$; after Tukey's HSD pairwise comparison test) (Figure II.3b). Overall, maleness of *A. artemisiifolia* also indicated a significant main effect of neighboring plant species, but this effect could not be shown with pairwise comparison tests (Tukey HSD; $P > 0.05$) (Supplemental Table A.II.2).

In terms of mycorrhization, plant roots in the pots treated with AM fungal inoculum were colonized by AM fungi: percentage colonization by AM hyphae

38.3 ± 5.7 %, arbuscules 12.7 ± 4.2 %, and vesicles 6.9 ± 1.0 %. Infection with non-AM fungi was very low (0.2 ± 0.2 %) in the mycorrhizal treatment. Non-mycorrhizal controls were also rarely infected by fungi (1.0 ± 0.3 %), none of which could be classified as AM fungi.

3.4.2 Pairwise competition experiment

I found that the roadside AM fungal community had divergent effects on performance of *A. artemisiifolia* and *D. carota* grown in pairwise competitive situations (significant main effect of mycorrhizal treatment; Table II.2). In the presence of AM fungi, *D. carota* produced always significantly more shoot and root biomass (Figure II.4a; Table II.3). Conversely, *A. artemisiifolia* had significantly greater shoot biomass without AM fungi; root biomass showed similar, but non-significant results (Figure II.4a; Table II.3). Furthermore, shoot biomass of both plant species was also divergently influenced by intraspecific and interspecific competition, respectively (significant main effect of competitive situation; Table II.2). While *D. carota* had greater biomass when grown with a conspecific, *A. artemisiifolia* produced significantly more biomass in the mixed situation (Figure II.4b).

Considering the influence of AM fungi on shoot and root biomass of *A. artemisiifolia* and *D. carota* in the different competitive situations, I found partially similar results as already indicated by the target–challenger experiment. When *A. artemisiifolia* grew with a conspecific, I found that its shoot and root biomass was significantly decreased with AM fungi compared to without (Table II.3). In mixture with *D. carota* and presence of AM fungi, *A. artemisiifolia* had also less biomass compared to the non-mycorrhizal control, although not significantly (Table II.3). A similar pattern, i.e., reduced shoot growth of *A. artemisiifolia* under condition of AM fungi and *D. carota* as competitor, was already found when *A. artemisiifolia* was target to *D. carota*. Regarding *D. carota* in mixture, shoot and root biomass was significantly increased in the presence of AM fungi compared to their absence (Table II.3). Such a positive mycorrhizal effect on *D. carota* was also already indicated by the target-challenger experiment, but results were non-significant (target arrangement with neighbor *D. carota;* shoot biomass with AM fungi: 0.546 ± 0.057 g; non-mycorrhizal control: 0.400 ± 0.042 g; Supplemental Table A.II.1). Furthermore, shoot and root biomass of *D. carota* in competition with a conspecific was also significantly increased in symbiosis with AM fungi than without (Table II.3).

Concerning mycorrhization, roots of both plant species were colonized by AM fungi when soil was treated with AM fungal inoculum. Overall, AM fungal colonization tended to be higher in roots of *D. carota* (mean ± SE; hyphae 44.7 ± 7.3 %; arbuscules 27.0 ± 5.1 %) than *A. artemisiifolia* (hyphae 36.2 ± 7.7 %; arbuscules 20.9 ± 5.1 %), but this difference was non-significant. Furthermore, percentage colonization by AM fungal hyphae, arbuscules and vesicles of both plant species showed no significant difference depending on competitive situation, although *A. artemisiifolia* tended to form more AM fungal structures in the roots when grown with *D. carota* (Supplemental Table A.II.5). Non-mycorrhizal controls were broadly similar infected with non-AM fungi (0.3 ± 0.1 %) as plants growing in AM fungal treatment (0.2 ± 0.1 %).

3.5 Discussion

In contrast to my first hypothesis, the AM fungal symbiosis with natural AM fungal communities was of minor importance for the competitive ability of *A. artemisiifolia* in target and challenger arrangements with co-existing ruderal mycorrhizal competitors. Regardless of presence/absence of AM fungi, *A. artemisiifolia* grew considerably taller compared to neighboring plant species in all performed situations of interspecific competition; hence, *A. artemisiifolia* was highly dominant under the low fertile soil conditions tested here. This result is supported by findings of Miller (1994), who also found *A. artemisiifolia* as an exceptionally good competitor that had strong direct suppressive effects on four co-existing plant species in North America. Moreover, Tilman (1986) showed that *A. artemisiifolia* had the greatest biomass as seedling when grown under low nitrogen levels in comparison to other eight co-occurring plant species. However, Leskovšek et al. (2012) reported that *A. artemisiifolia* was a poor competitor in competition with *Lolium multiflorum* L. under high resource availability (high levels of nitrogen and water). Further, they demonstrated that growth of *A. artemisiifolia* in competition was minimally affected by shortage of nutrients. Therefore, one explanation for the competitive dominance of *A. artemisiifolia* in my study may be its ability to compete for nitrogen in nutrient poor soils, such as those used here.

In accordance to my second hypothesis, growth of the neighboring plants was more decreased when they where surrounded by *A. artemisiifolia* (high relative density of *A. artemisiifolia*). The growth of all species strongly differed in target compared to challenger arrangements of *A. artemisiifolia*. Interestingly, identity of neighbor species had a significant impact on shoot biomass both of *A. artemisiifolia* and the neighboring species grown in target or challenger situations, which was contrary to my third hypothesis and other observations. For example, Miller (1994) found no evidence that growth of *A. artemisiifolia* was differently affected by presence of any other species. However, for other invasive plants such as *Centaurea melitensis* or *Centaurea stoebe* (formerly *C. maculosa*) it has been shown that plant neighbor identity matters (Callaway et al. 2003; 2004).

Considering the neighbors in my study, the annual *C. canadensis* performed poorly in all arrangements. The other species, i.e., *D. carota*, *A. vulgaris* or *T. vulgare*, grew equally poor in challenger arrangements, but performed differently when *A. artemisiifolia* was the target. In target situations of *A. artemisiifolia*, the upright perennials *A. vulgaris* and *T. vulgare* had on average the greatest shoot biomass of the neighboring species selected, while the biennial *D. carota* grew less compared to *A. vulgaris*. Conversely, for target *A. artemisiifolia* I found that its shoot growth differed in the presence of the most inferior competitor *C. canadensis* only. Here, target *A. artemisiifolia* could profit most, and had almost doubled its shoot biomass compared to arrangements with *D. carota*, *A. vulgaris* and *T. vulgare*. However, the substantial competitive advantage of *A. artemisiifolia* over *C. canadensis* may have been related to their different growth forms. During the whole period of the experiment, *C. canadensis* was in the leaf rosette-forming stage and, therefore, additionally shaded by leafs of the upright growing *A. artemisiifolia*.

In challenger arrangements, i.e., when *A. artemisiifolia* grew also with conspecifics, its shoot growth was strongly decreased compared to target arrangements irrespective of the neighboring plant and presence/absence of mycorrhiza, respectively. Since shoot performance of neighbors was also strongly suppressed when surrounded by *A. artemisiifolia*, I interpret my findings of reduced growth of challenger *A. artemisiifolia* related to a strong intraspecific competition within *A. artemisiifolia*. As shown by other studies, performance of *A. artemisiifolia* is reduced with increasing density of conspecifics (Miller and Werner 1987; MacDonald and Kotanen 2010). In the experiment, therefore, it seems that challenger plants of *A. artemisiifolia* were strongly competing with each other, which may have overridden other effects, such as the impact of mycorrhizal fungi.

Considering the mycorrhizal impact on *A. artemisiifolia* as target, I found that the AM fungal effect pointed towards different directions depending on the neighboring plant species, although not significantly. AM fungi tended to have a positive effect on target *A. artemisiifolia* when competing with *C. canadensis* or *T. vulgare*. In competition with *A. vulgaris*, however, the mycorrhizal symbiosis mediated a neutral effect on *A. artemisiifolia*. When *A. artemisiifolia* was surrounded by *D. carota*, shoot growth of *A. artemisiifolia* was reduced with AM fungi compared to without; here, the mycorrhizal symbiosis amplified effects of interspecific competition.

This growth reducing effect of AM fungi on *A. artemisiifolia* in the presence of *D. carota* was also indicated by the pairwise competition experiment solely focusing on *A. artemisiifolia* and *D. carota* in intra- and interspecific competitive situations. Again, I found that *A. artemisiifolia* tended to produce less biomass when growing in mixture with *D. carota* under mycorrhizal compared to non-mycorrhizal conditions.

However, when *A. artemisiifolia* was tested in intraspecific competition I even found a stronger negative effect of AM fungi on competing conspecifics; hence, the mycorrhizal symbiosis clearly amplified competition within *A. artemisiifolia*, which is in line with recent findings of a meta-analysis by Moora and Zobel (2010). They showed that AM fungi have an amplifying or neutral effect in intraspecific competition, while the effect is balancing in interspecific competition. Since I maintained the ecological context of soil, AM fungal community and plant species origin in my study, the AM fungal effect on *A. artemisiifolia* under natural conditions must be assumed to be negative in intraspecific competition. The effect in interspecific competition with *D. carota* might be rather neutral (because biomass was not significantly reduced). The findings of my pairwise competition experiment are contrary to a study by Shah et al. (2008) investigating the invasive plants *Anthemis cotula* in India. Similarly, they studied the effect of AM fungi on *A. cotula* and *D. carota* in intra- and interspecific competition. Here, invasive *A. cotula* was enhanced by the presence of local AM fungi when grown together with *D. carota*. Further, they found that *A. cotula* was even more promoted by AM fungi when grown in monoculture, which is the exact opposite of what I found for *A. artemisiifolia*. Shah et al. (2008) also showed that the degree of AM fungal root colonization of *A. cotula* was decreased in mixture with *D. carota* compared to monoculture. My data, however, indicate that roots of *A. artemisiifolia* were more strongly colonized by AM fungi when grown in mixture with *D. carota*. In my experiment, furthermore, *D. carota* strongly profited from the mycorrhizal symbiosis both in intra- and interspecific competition; thus, the AM fungal symbiosis had a balancing effect on intraspecific competition of *D. carota*, which is in contrast to *A. artemisiifolia* and other plant species (Moora and Zobel 2010). Moreover, the competitive ability of *D. carota* was increased in interspecific competition with *A. artemisiifolia* in the presence of the roadside AM fungal assemblage tested here. Aside from AM fungi, *A. artemisiifolia* was most strongly affected by conspecific competition in both experiments. Conversely, *D. carota* performed better in intraspecific competition.

Considering reproductive traits of *A. artemisiifolia* in the target–challenger experiment, I found that the number of male inflorescences was significantly higher when it grew as target to the most inferior competitor *C. canadensis* under mycorrhizal compared to non-mycorrhizal conditions. Interestingly, increased production of male flowers coincided with greatest shoot biomass produced by *A. artemisiifolia*. Such positive effects of mycorrhizal symbiosis on shoot biomass and maleness were also found by Koide and Li (1991). Furthermore, *A. artemisiifolia* also showed protandry in specific combinations of interspecific competition: as target to *C. canadensis* (mycorrhizal and non-mycorrhizal treatment) and *T. vulgare* (only mycorrhizal). The phenomenon that sex allocation in *A. artemisiifolia* is adjusted to size and environmental conditions has been recently reported by Friedman and Barrett (2011). They showed that flowering can range from protandry in the sun to protogyny in the shade. Ackerly and Jasieński (1990), moreover, demonstrated that the variability in aboveground biomass and gender is higher under nutrient-rich soil conditions leading to a shift towards maleness in taller plants. Similarly, other studies also found that favorable conditions increase maleness in *A. artemisiifolia* (McKone and Tonkyn 1986; Traveset 1992; Bäucker, present book, Study I). Therefore, my data once more demonstrate that greater shoot biomass in *A. artemisiifolia* results in higher number of male inflorescences.

Regarding female function of *A. artemisiifolia*, I found that number of female flowers was increased more than five-fold when *A. artemisiifolia* had a high relative density (grown as challenger) compared to low (target). Thus, my data suggest that reduced shoot growth and lower production of male flowers in competition with conspecifics led to earlier development of female flowers in *A. artemisiifolia*. A study by Lundholm and Aarssen (1994) also reported increased female function of *A. artemisiifolia* in the presence of neighboring plants. However, an earlier flowering of female flowers in *A. artemisiifolia* does not necessarily imply greater seed output. There are some studies showing that plants, which produce less vegetative biomass also have lower seed mass (Chikoye et al. 1995; Leskovšek et al. 2012; Bäucker, present book, Study I).

To conclude, *A. artemisiifolia* was demonstrated to be an exceptionally good competitor at low and high relative density in comparison to co-occurring mycorrhizal plant species in the new European range under nutrient poor soil conditions. Moreover, *A. artemisiifolia* experienced strong competition by conspecifics, which caused decreases in shoot biomass and maleness, but earlier flowering of female flowers. Further, I found no evidence that growth performance or competitive ability of *A. artemisiifolia* was enhanced in the presence of natural AM fungal communities under conditions of a maintained ecological context. In fact, my findings show that AM fungi had an amplifying effect on *A. artemisiifolia* in pairwise intraspecific competition, and a neutral effect in mixture with *D. carota*. Therefore, *A. artemisiifolia*, a successful pioneer plant and a species with a strongly ruderal life history, has a weak dependence on symbiosis with AM fungi. Therefore, the invasive success of *A. artemisiifolia* in Central Europe may not be related to facilitation by natural AM fungal communities, as previously proposed by Fumanal et al. (2006). However, since I studied the competitive ability of *A. artemisiifolia* and co-existing plant species at the seedlings stage, the outcome of interspecific competition may change as the species grow longer at a site under mycorrhizal conditions, which needs further research.

3.6 References

Ackerly DD, Jasieński M (1990) Size-dependent variation of gender in high density stands of the monoecious annual, *Ambrosia artemisiifolia* (Asteraceae). Oecologia 82:474-477.

Bazzaz FA, Mezga DM (1973) Primary productivity and microenvironment in an Ambrosia-dominated old field. Am Midl Nat 90:70-78.

Brandes D, Nitzsche J (2006) Biology, introduction, dispersal, and distribution of common ragweed (*Ambrosia artemisiifolia* L.) with special regard to Germany. Nachr.bl Dtsch Pflanzenschutzd 58:286-291.

Brandes D, Nitzsche J (2007) Verbreitung, Ökologie und Soziologie von *Ambrosia artemisiifolia* L. in Mitteleuropa. Tuexenia 27:167-194.

Bray SR, Kitajima K, Sylvia DM (2003) Mycorrhizae differentially alter growth, physiology, and competitive ability of an invasive shrub. Ecol Appl 13:565-574.

Callaway RM, Mahall BE, Wicks C, Pankey J, Zabinski C (2003) Soil fungi and the effects of an invasive forb on grasses: neighbour identity matters. Ecology 84:129-135.

Callaway RM, Newingham B, Zabinski CA, Mahall BE (2005) Compensatory growth and competitive ability of an invasive weed are enhanced by soil fungi and native neighbours. Ecol Lett 4:429-433.

Callaway RM, Thelen GC, Barth S, Ramsey PW, Gannon JE (2004) Soil fungi interactions between the invader *Centaurea maculosa* and North American natives. Ecology 85:1062-1071.

Chikoye D, Weise SF, Swanton CF (1995) Influence of common ragweed (*Ambrosia artemisiifolia*) time of emergenze and density on white bean (*Phaseolus vulgaris*). Weed Science 43:375-380.

Crowell HF, Boerner REJ (1988) Influences of mycorrhizae and phosphorus on belowground competition between two old-field annuals. Environ Exper Bot 28:381-392.

Ellmer F, Peschke H, Köhn W, Chmielewski FM, Baumecker M (2000) Tillage and fertilizing effects on sandy soils. Review and selected results of long-term experiments at Humboldt-University Berlin. J. Plant Nutr Soil Sci 163:267-272.

Emery SM, Rudgers JA (2012) Impact of competition and mycorrhizal fungi on growth of *Centaurea stoebe,* an invasive plant of sand dunes. Am Midl Nat 167:213-222.

Friedman J, Barrett SCH (2011) Genetic and environmental control of temporal and size-dependent sex allocation in a wind-pollinated plant. Evolution 65:2061-2074.

Fumanal B, Plenchette C, Chauvel B, Bretagnolle F (2006) Which role can arbuscular mycorrhizal fungi play in the facilitation of *Ambrosia artemisiifolia* L. invasion in France? Mycorrhiza 17:25-35.

Helgason T, Fitter AH (2009) Natural selection and the evolutionary ecology of the arbuscular mycorrhizal fungi (Phylum Glomeromycota). J Exp Bot 60:2465-2480.

Hawkes CV, Belnap J, D'Antonio C, Firestone MK (2006). Arbuscular mycorrhizal assemblages in native plant roots change in the presence of invasive exotic grasses. Plant Soil 281:369-380.

Hoeksema JD, Chaudhary VB, Gehring, CA, Johnson NC, Karst J, Koide RT, Pringle A, Zabinski C, Bever JD, Moore JC, Wilson GWT, Klironomos JN, Umbanhowar J (2010) A meta-analysis of context-dependency in plant response to inoculation with mycorrhizal fungi. Ecol Lett 13:394-407.

Jeschke JM, Gómez Aparicio L, Haider S, Heger T, Lortie CJ, Pyšek P, Strayer DL (2012) Support for major hypotheses in invasion biology is uneven and declining. NeoBiota 14:1-20.

Johnson NC, O'Dell TE, Bledsoe CS (1999) Methods for ecological studies of mycorrhizae. In: Robertson GP, Coleman DC, Bledsoe CS, Sollin P (eds) Standard soil methods for long-term ecological research, Oxford University Press, New York, pp. 378-412.

Keller RP, Geist J, Jeschke JM, Kühn I (2011) Invasive species in Europe: ecology, status, and policy. Environmental Sciences Europe 23:23.

Koide RT, Li M (1991) Mycorrhizal fungi and the nutrient ecology of three oldfield annual plant species. Oecologia 85:403-412.

Leskovšek R, Eler K, Batič F, Simončic A (2012) The influence of nitrogen, water and competition on the vegetative and reproductive growth of common ragweed (*Ambrosia artemisiifolia* L.). Plant Ecol 213:769-781.

Lundholm JT, Aarssen LW (1994) Neighbor effects on gender variation in *Ambrosia artemisiifolia*. Can J Bot 72:794-800.

MacDonald AAM, Kotanen PM (2010) The effects of disturbance and enemy exclusion on performance of an invasive species, common ragweed, in its native range. Oecologia 162:977-986.

Marler MJ, Zabinski CA, Callaway RM (1999) Mycorrhizae indirectly enhance competitive effects of an invasive forb on a native bunchgrass. Ecology 80:1180-1186.

McGonigle TP, Miller MH, Evans DG, Fairchild GL, Swan JA (1990) A new method which gives an objective measure of colonization of roots by vesicular arbuscular mycorrhizal fungi. New Phytol 115:495-501.

McKone MJ, Tonkyn DW (1986) Intrapopulation gender variation in common ragweed (Asteraceae, *Ambrosia artemisiifolia* L.), a monoecious, annual herb. Oecologia 70:63-67.

Meinhardt KA, Gehring CA (2012) Disrupting mycorrhizal mutualisms: a potential mechanism by which exotic tamarisk outcompetes native cottonwoods. Ecol Appl 22:532-549.

Miller TE (1994) Direct and indirect species interactions in an early old-field plant community. Am Nat 143:1007-1025.

Miller TE, Werner PA (1987) Competitive effects and responses between plant species in a first-year old-field community. Ecology 68:1201-1210.

Moora M, Zobel M (2010) Arbuscular mycorrhizae and plant–plant interactions: impact of invisible world on visible patterns. In: Pugnaire FI (ed) Positive plant interactions and community dynamics, CRC press, New York, pp. 79-98.

Mummey DL, Rillig MC (2006) The invasive plant species *Centaurea maculosa* alters arbuscular mycorrhizal fungal communities in the field. Plant Soil 288:81-90.

Newsham KK, Fitter AH, Watkinson AR (1995) Multi-functionality and biodiversity in arbuscular mycorrhizas. Trends Ecol Evol 10:407-411.

Pejchar L, Mooney HA (2009) Invasive species, ecosystem services and human well-being. Trends Ecol Evol 24:497-504.

Pimentel D, Zuniga R, Monison D (2005) Update on the environmental and economic costs associated with alien-invasive species in the United States. Ecol Econ 52:273-288.

Pringle A, Bever JD, Gardes M, Parrent JL, Rillig MC, Klironomos JN (2009) Mycorrhizal symbioses and plant invasions. Annu Rev Ecol Evol S 40:699-715.

R Development Core Team (2009) R: A language and environment for statistical computing. R Foundation for Statistical Computing, Vienna, Austria. ISBN 3-900051-07-0, URL http://www.R-project.org.

Rebele F (2000) Competition and coexistence of rhizomatous perennial plants along a nutrient gradient. Plant Ecol 147:77-94.

Reinhart KO, Callaway RM (2006) Soil biota and invasive plants. New Phytol 170:445-457.

Richardson DM, Allsopp N, D'Antonio CM, Milton SJ, Rejmánek M (2000b) Plant invasions – the role of mutualisms. Biol Rev Camb Philos 75:65-93.

Richardson DM, Pyšek P, Rejmanek M, Barbour MG, Panetta FD, West CJ (2000a) Naturalization and invasion of alien plants: concepts and definitions. Divers Distri 6:93-107.

Scheublin TR, van Logtestijn RSP, van der Heijden MAG (2007) Presence and identity of arbuscular mycorrhizal fungi influence competitive interactions between plant species. J Ecol 95:631-638.

Shah MA, Reshi Z, Reshi I (2008) Mycorrhizosphere mediated mayweed chamomile invasion in the Kashmir Himalaya, India. Plant Soil 312:219-225.

Shah MA, Reshi ZA, Khasa DP (2009) Arbuscular mycorrhizas: drivers or passengers of alien plant invasion. Bot Rev 75:397-417.

Stinson KA, Campbell SA, Powell JR, Wolfe BE, Callaway RM, Thelen GC, Hallett SG, Prati D, Klironomos JN (2006) Invasive plant suppresses the growth of native tree seedlings by disrupting belowground mutualisms. PloS Biol 4:1-5.

Tilman D (1986) Nitrogen-limited growth in plants from different successional stages. Ecology 67:555-563.

Traveset A (1992) Sex expression in a natural population of the monoecious annual, *Ambrosia artemisiifolia* (Asteraceae). Am Midl Nat 127:309-315.

Vierheilig H, Coughlan AP, Wyss U, Piche Y (1998) Ink and vinegar, a simple staining technique for arbuscular-mycorrhizal fungi. Appl Environ Microb 12:5004-5007.

Vilà M, Basnou C, Pyšek P, Josefsson M, Genovesi P, Gollasch S, Nentwig W, Olenin S, Roques A, Roy D, Hulme PE, and DAISIE partners (2010) How well do we understand the impacts of alien species on ecosystem services? A pan-European, cross-taxa assessment. Front Ecol Environ 8:135-144.

Wardle DA, Bardgett RD, Callaway RM, van der Putten WH (2011) Terrestrial ecosystem responses to species gains and losses. Science 332:1273-1277.

Wilson GWT, Hickman KR, Williamson MM (2012) Invasive warm-season grasses reduce mycorrhizal root colonization and biomass production of native prairie grasses. Mycorrhiza 22:327-336.

Zabinski CA, Quinn L, Callaway RM (2002) Phosphorus uptake, not carbon transfer, explains arbuscular mycorrhizal enhancement of *Centaurea maculosa* in the presence of native grassland species. Funct Ecol 16:758-765.

Zhang Q, Yao LJ, Yang RY, Yang XY, Tang JJ, Chen X (2007) Potential allelopathic effects of an invasive species *Solidago canadensis* on the mycorrhizae of native plant species. Allelopathy J 20:71-78.

3.7 Tables and figures

Table II.1 Analyses of variance (ANOVAs) on response variables of *Ambrosia artemisiifolia* and neighboring plant species (Neighbor) in the target–challenger experiment, with D (relative density of *A. artemisiifolia*), Sp (neighboring plant species) and Treat (soil treatment) as factors. Values in bold indicate significance at $P < 0.05$. Overview on analyzed data is given in Supplemental Table A.II.1.

		Ambrosia artemisiifolia						Neighbor	
		Shoot biomass		No. female flowers		No. male flowers		Shoot biomass	
Factors	df	F	P	F	P	F	P	F	P
D	1	**490.1**	**<0.001**	**86.8**	**<0.001**	**18.6**	**<0.001**	**152.5**	**<0.001**
Sp	3	**15.8**	**<0.001**	0.6	0.641	**3.4**	**0.022**	**43.3**	**<0.001**
Treat	1	0.6	0.600	2.7	0.103	2.8	0.100	1.6	0.212
D x Sp	3	**16.5**	**<0.001**	2.4	0.076	1.5	0.214	**6.2**	**<0.001**
D x Treat	1	0.9	0.359	1.0	0.312	1.5	0.231	0.6	0.455
Sp x Treat	3	**4.6**	**0.005**	1.0	0.396	2.2	0.092	0.5	0.654
D x Sp x Treat	3	**3.4**	**0.022**	0.1	0.977	**2.9**	**0.038**	1.6	0.193
Residuals			95		95		95		95

Table II.2 Analyses of variance (ANOVAs) on biomass traits of *Ambrosia artemisiifolia* (A) and *Daucus carota* (D) in the pairwise competition experiment in intraspecific (AA, DD) and interspecific competition (AD), with Comp (competitive situation) and Treat (soil treatment) as factors. Values in bold indicate significance at $P < 0.05$; analyzed data are shown Table II.3.

| | | Shoot biomass | | | |
| | | *A. artemisiifolia* in AA vs. AD | | *D. carota* in DD vs. AD | |
Factors	df	F	P	F	P
Comp	1	**5.64**	**0.022**	**4.10**	**0.049**
Treat	1	**5.26**	**0.027**	49.08	**<0.001**
Comp x Treat	1	0.04	0.851	0.30	0.584
		Root biomass			
		A. artemisiifolia in AA vs. AD		*D. carota* in DD vs. AD	
Factors	df	F	P	F	P
Comp	1	1.09	0.302	1.58	0.216
Treat	1	1.33	0.255	**24.42**	**<0.001**
Comp x Treat	1	1.09	0.303	0.57	0.456
Residuals		44		44	

Table II.3 Results of paired *t*-test on shoot and root biomass of *Ambrosia artemisiifolia* (A) and *Daucus carota* (D) in intra- and interspecific competition, and in the presence or absence of AM fungi. In intraspecific competition, data (mean ± SE) represent the mean of the two conspecifics growing together. *P*-values in bold indicate significance at $P < 0.05$.

	Soil treatment		*t*-test	
	AM fungi	non-mycorr.	df	*P*-value
intraspecific competition				
Shoot biomass A (g)	108.4 ± 9.5	133.9 ± 6.0	11	**0.040**
Root biomass A (g)	39.8 ± 7.1	53.2 ± 5.2	11	**0.015**
Shoot biomass D (g)	38.1 ± 5.4	8.1 ± 1.2	11	**<0.001**
Root biomass D (g)	14.5 ± 2.8	4.5 ± 0.4	11	**0.003**
interspecific competition				
Shoot biomass A (g)	140.8 ± 13.1	179.8 ± 19.4	11	0.191
Root biomass A (g)	36.6 ± 5.2	49.2 ± 10.8	11	0.394
Shoot biomass D (g)	26.5 ± 5.2	5.9 ± 0.8	11	**0.002**
Root biomass D (g)	10.5 ± 1.9	4.1 ± 0.3	11	**0.010**

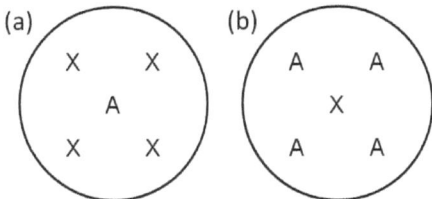

Figure II.1 Target versus challenger arrangement of *Ambrosia artemisiifolia* (A) in the target–challenger experiment: (a) shows *A. artemisiifolia* growing as target (relative density 1:4), and (b) as challenger (relative density 4:1) to the neighboring plant species (X).

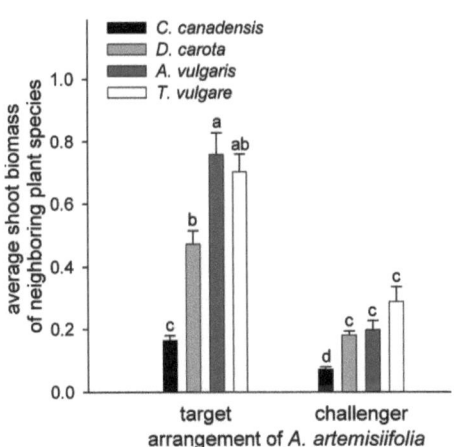

Figure II.2 Shoot biomass of the four neighboring plant species in response to relative density of *Ambrosia artemisiifolia* in the experiment (black = *Conyza canadensis*, light grey = *Daucus carota*; dark grey = *Artemisia vulgaris*, white = *Tanacetum vulgare*). Bars show means ± SE, and different lower case letters on the bars indicate significant differences ($P < 0.05$) among treatment groups according to Tukey's HSD test.

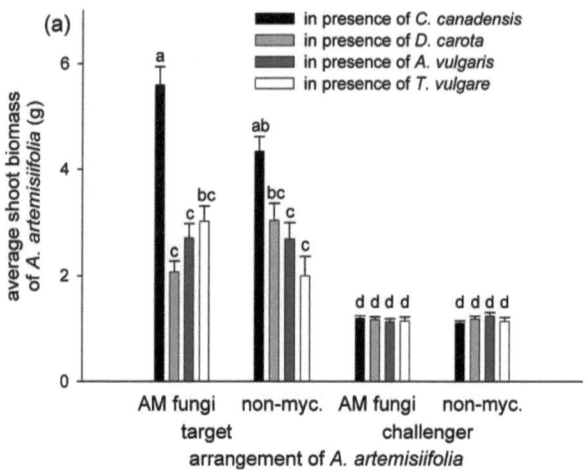
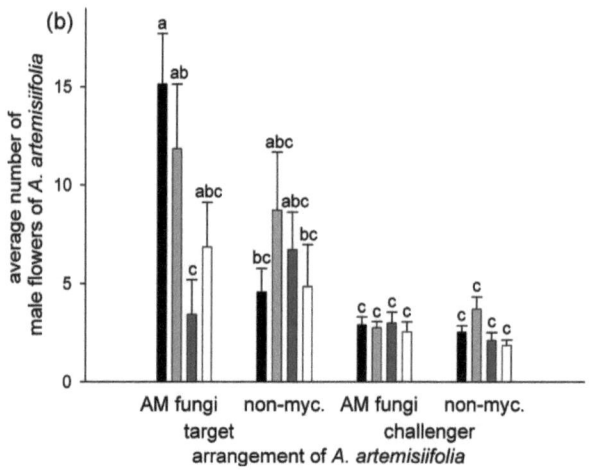

Figure II.3 Shoot biomass (a) and number of male inflorescences (b) of *Ambrosia artemisiifolia* in response to its relative density, soil treatment (mycorrhizal vs. non-mycorrhizal) and the four neighboring plant species (black = *Conyza canadensis*, light grey = *Daucus carota*; dark grey = *Artemisia vulgaris*, white = *Tanacetum vulgare*). Bars represent means ± SE. Different lower case letters on the bars indicate significant differences ($P < 0.05$) among treatment groups according to Tukey's HSD test.

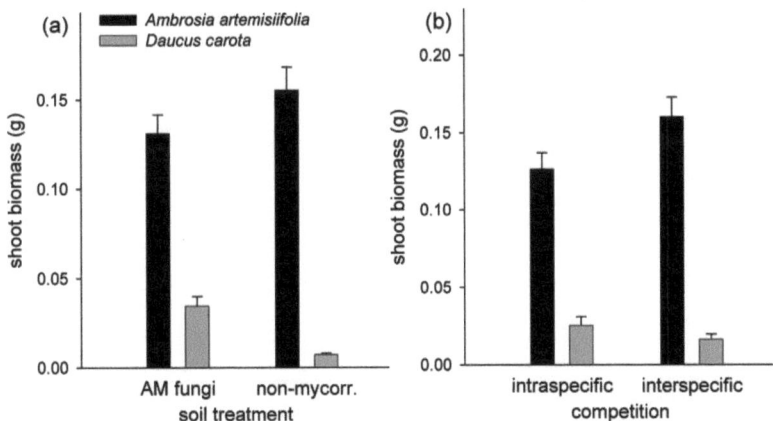

Figure II.4 Shoot biomass of *Ambrosia artemisiifolia* (black) and *Daucus carota* (grey) in response to the main effects soil treatment (a) and competitive situation (b). In (b) data for intraspecific arrangements were calculated from averaged values of both plants per pot. Bars represent means ± SE.

4 STUDY III: DISTINCT SEED MORPHS OF *GALINSOGA PARVIFLORA* (ASTERACEAE) GIVE RISE TO DIFFERENT SOIL FEEDBACKS

4.1 Abstract

Heterocarpy is the phenomenon that a single plant produces two or more distinct fruit types, which differ in dispersal mechanisms and ecological behavior. *Galinsoga parviflora* represents such a heterocarpic plant that produces capitula with two different seed morphs: seeds equipped with a pappus for long-distance dispersal and non-pappus seeds, which have a low dispersal potential. I studied if plants arising from the two seed types differ in their soil feedback responses. Soil feedback is known to affect plant performance often negatively in trained ('self-cultivated') soil due to accumulation of disadvantageous soil organisms. In the experiment, I trained soil with plants arising from either non-pappus or pappus seeds over two plant generations, and tested soil feedback responses of non-pappus and pappus progeny as 'trained versus sterile' soil contrasts. I found that plants from pappus seeds, which are produced to colonize new sites, suffered greater negative feedback than plants from non-pappus seeds, which are produced for *in situ* persistence. The soil feedback differences were most pronounced for reproductive traits, but also indicated by root biomass. Moreover, non-pappus progeny produced a higher portion of pappus seeds, i.e. showed greater investment in dispersal in adverse conditions of trained soil compared to pappus progeny. The soil feedback differences between non-pappus and pappus progeny could not be shown to be related to differences in fungal root colonization; hence, different magnitudes of negative soil feedback must have been caused by soil organisms other than fungi. Interestingly, pappus progeny produced significantly more root biomass when they came from non-pappus mother/grandmother plants. Such maternal effects need further evaluation and might be more common in heterocarpic plants.

4.2 Introduction

Many plant species have constant seed size (Harper et al. 1970), while other species show intra-individual variation in seed form or behavior, termed seed heteromorphism (Venable 1985a). Such heteromorphism becomes most evident when two or more distinctly different fruit types with divergent ecological functions are produced by one plant. This phenomenon was described as heterocarpy (Mandák 1997), and it is often associated with differential dispersal mechanisms and differential germination responses of the respective fruit types (Tanowitz et al. 1987). Beside fruit, also the terms diaspore (i.e. dispersal unit) or seed in the broadest sense are used in the literature: in this study I refer to seed *sensu lato*.

Within the angiosperms, heterocarpy/heterospermy is frequently known in Asteraceae and Chenopodiaceae (Imbert 2002). Predominantly, heterocarpy occurs among annuals, pioneer species or plants of stochastic environments, such as deserts or semi-deserts (Mandák 1997). It represents an adaptive trait evolving in situations of environmental unpredictability (Cruz-Mazo et al. 2009) or when the evolution of plasticity is partially impeded (Simons 2011). The most noticeable feature of heterocarpic plants is the diversification of offspring in space or time, such as non-dispersive vs. dispersive or dormant vs. non-dormant seed types (Venable et al. 1995). In this context, dormancy has been found to be negatively correlated with dispersal (Venable and Brown 1988; Venable 1989): higher dormancy results in reduced selection for dispersal and vice versa (Venable et al. 1995).

The advantage of producing different kind of seeds per individuum under environmental unpredictability has often been considered from the perspective of bet-hedging, i.e. trading-off some potential short-term benefit for a long-term benefit (e.g. Venable 1985a; Philippi and Seger 1989; Venable 2007). Such a strategy ensures that at least some offspring successfully cope with conditions of an unpredictable environment (Venable et al. 1995). As a consequence, bet-hedging traits do not maximize the expected fitness within a generation, but do maximize geometric-mean fitness across generations, which results in higher long-term success of bet hedgers compared to non-bet hedgers (Simons 2011). Other explanations for the evolution of heterocarpy are escape from negative density effects or sib competition (Venable and Brown 1988).

A large body of literature exists on differences of seed morphs of heterocarpic plants, where patterns of seed size, dormancy and germination have extensively been studied. For Asteraceae, many studies indicate that achenes produced by central flowers of the capitula have lower dormancy and their germination is less restricted by specific temperature regimes (Baskin and Baskin 1976; Flint and Palmblad 1978; McEvoy 1984; Tanowitz et al. 1987; Venable and Levin 1985a; De Clavijo 2001; Brändel 2007; Sun et al. 2009; Aguado et. al 2011; but see for the opposite pattern Brändel 2004; Rai and Tripathi 1982). Moreover, the central achenes were frequently found to be lighter than the peripheral ones, although also some other patterns were reported for seed mass and/or embryo weight (Rocha 1996; Brändel 2004; Venable and Levin 1985b). In a few species, however, heterocarpy is not associated with germination differences (Baker and O'Dowd 1982; Sorensen 1978; Imbert et al. 1996); hence, different seed morphologies reflect different seed dispersal strategies and, other differences in ecological behavior.

Therefore, the influence of environmental conditions on heterocarpic plant traits, such as the ratio of non-dispersing to dispersing seeds has occasionally been studied (e.g. Ellner and Schmida 1984; Imbert and Ronce 2001; Kigel 1992; Cheptou et al. 2008). For example, Cheptou et al. (2008) demonstrated that *Crepis sancta* responds to urban habitat fragmentation with a shift towards a higher portion of seeds lacking dispersal structures, which reduces costs of dispersal. In a few other studies, the effects of density/competition and soil conditions, like nutrient or water availability, have been tested by comparing plants arising from the distinct seed morphs. For Asteraceae, some studies report differences in competitive ability of plants originating from different seed types (Venable 1985b; Rai and Tripathi 1987; Imbert et al. 1997; De Clavijo and Jiménez 1998; but see Sorensen 1978; Baker and O'Dowd 1982; Brändel 2007). In Chenopodiaceae, plants from different seed types were also shown to differ in biomass

(Ellison 1987), as well as resource allocation to the different seed types when grown both in mixture with each other and monoculture (Mandák and Pyšek 2005).

From the existing literature it can be seen that the phenomenon of heterocarpy is very complex. Since heterocarpy was predominantly investigated from the perspective of the different seeds, my aim was to study another ecological aspect. I tested if plants arising from the different seed types differ in their interaction with soil biota; hence, show divergent responses of soil feedback. In short, soil feedback is known to affect plant performance often negatively in trained ('self-cultivated') soil due to accumulation of disadvantageous soil organisms, such as soil-borne bacteria, fungi and invertebrate fauna (Bever 1994; Klironomos 2002; Bever 2002; De Deyn et al. 2003). The negative soil feedback effect becomes visible when a plant species grows less in its trained soil compared to soils trained with other plant species. Negative soil feedback is known to be an important local and large-scale mechanism influencing plant abundance and mediating plant coexistence (Bever 2003; Ehrenfeld et al. 2005; Kulmatiski et al. 2008; Petermann et al. 2008; Johnson et al. 2012).

To test for a relationship between dispersal ability and soil feedback I studied *Galinsoga parviflora* Cav. (Asteraceae). The species produces different floret types within a flower head (capitulum), where a group of bisexual disc florets in the center of the capitulum is surrounded by a ring of a few ray florets, which are female (Nielreich 1866). The different florets develop to distinctly different achene types: disc achenes equipped with a pappus and ray achenes lacking a pappus as illustrated by Becker (1913). The ripe ray achenes (hereafter non-pappus seeds) remain enclosed in involucral bracts forming a winged structure (Espinosa-García and Sarukhán 1997), which may disperse or fall as a whole into the local habitat. Because non-pappus seeds lack visible adaptation to long-distance dispersal (Vibrans 1999), they are produced for short-distances dispersal. In contrast, disc achenes (hereafter pappus seeds) possess a crown of scales as appendages best-suited for long-distance dispersal predominately by human clothing (Holm et al. 1977; Vibrans 1999), but also by animal fur (Vibrans 1999) or wind (Terzioğlu and Anşin 2001).

Besides dispersal structures, the two seed types of *G. parviflora* differ in other aspects. Under low and high light regimes, non-pappus seeds germinate earlier and at higher percentages compared to pappus seeds (Rai and Tripathi 1982, 1987). The higher germination rate of non-pappus seeds is probably due to their greater seed weight and higher contents of reserves in comparison to pappus seeds (Rai and Tripathi 1982). The two seed types, moreover, differ in their dormant characters, and thus loss rates from the seed bank (Espinosa-García et al. 2003). Further, non-pappus and pappus progeny show differences in seedling survival, as well as competiveness (Rai and Tripathi 1987).

Based on morphological and ecological differences of the two seed morphs of *G. parviflora,* it can be assumed that this species produces two functionally distinct seed types: non-pappus seeds for *in situ* persistence, and pappus seeds for colonizing new habitats; hence, founding of populations. Assuming that environments of newly colonized sites are highly dissimilar to existing populations, especially with regard to soil biota composition and abundance, the different dispersal capacities of seeds in *G. parviflora* may correspond to different soil feedback responses of plants arising from the different seed types. By implication, I hypothesized that plants grown from non-pappus seeds would exhibit better performance in trained soil than those from pappus seeds, i.e. progeny of non-pappus seeds experience less negative soil feedback compared to progeny of pappus seed.

4.3 Materials and methods

4.3.1 Study species

Galinsoga parviflora is an annual plant native to the mountainous region of Central America (Canne 1977). Several decades ago, the herb was already reported to have a worldwide distribution (Canne 1977; Holm et al. 1977), with human activity representing the most important vector (Warwick and Sweet 1983; Damalas 2008). The species occurs in disturbed habitats and agricultural areas, where it is an important weed (e.g. Holm et al. 1977; Warwick and Sweet 1983; Damalas 2008). *G. parviflora* reproduces via cross- and self-fertilization; hence, it needs one single seed only to start a new population (Warwick and Sweet 1983).

Seeds used in the experiment were collected in Warendorf Müssingen, Germany (51°58'06 N, 7°53'29 E) from plants growing on agricultural land. I ordered the seeds from the catalogue Index Seminum 2009 of the botanical garden of the Universität Münster (IPEN DE-0-MSTR-SA 8629).

4.3.2 Soils and soil preparation

The soils used for the experiment came from two different locations: Berlin-Dahlem (52°27' N, 13°18' E) and Thyrow (52°15' N, 13°14' E). The first site is an experimental field of the Institute of Biology of Freie Universität Berlin, the second belongs to a field research station of the Humboldt Universität Berlin. The urban Dahlem soil is classified as silty sand soil, while the Thyrow soil has a higher content of sand; hence, is characterized as sandy to silty sand soil (Ellmer et al. 2000). Further, Dahlem soil has a higher soil pH (pH = 6.1), higher contents of organic carbon (C = 1.01 %) and total nitrogen (N = 0.09 %) compared to Thyrow soil (pH = 5.2, C = 0.52 %, N = 0.04 %) (Schweitzer 2010; Baumecker et al. 2009).

Soil samples were taken from the top 15 cm of soil and sieved through a 4 mm sieve. Half of each soil was directly used for the training phase. The other half was stored (4 °C), and later autoclaved two times at 121 °C (each time 30 min) to prepare both sterile background and control soil, which were used in the final feedback experiment. Because the autoclaved soils had poor drainage, I mixed them with sterilized playground sand (ratio 5:1).

To prepare the soil treatment 'soil inoculated with trained soil', I thoroughly mixed trained soil into sterile background soil at the ratio of 1:10. The identity of replicates was always maintained so that seeds and soil inoculum used for testing soil-feedback responses had the same training phase context (Figure III.1).

4.3.3 Experiment

The experiment was based on the conceptual framework of plant-soil feedback (Bever 2003; Ehrenfeld et al. 2005). At first, it included a soil training phase, which was performed over two plant generations and followed by the final feedback experiment (Figure III.1). The feedback experiment had a fully 2 x 2 x 2 x 2 factorial design. It consisted of all combinations of soils from two locations (Thyrow soil, Dahlem soil), two seed morph histories resulting from the training phases (non-pappus seed history, pappus seed history), two seed types tested (non-pappus seed, pappus seed), and two soil treatments (soil inoculated with trained soil, sterile soil as control). During the training phase, I had four replicates for the different combinations of soil and seed type history. At the feedback stage, I replicated three times at the genotype level, which allowed me to account for the genotype effect. In the final experiment, I had 192 experimental units, where one individuum (plant originating from pappus seed type with pappus seed history in sterile Dahlem soil) died within the duration of the experiment.

For the soil training, I let plants grow from either non-pappus or pappus seeds in the two different soil types. After five weeks of growth, plants started to flower. Flowering stems with unopened capitula were separately enclosed in paper bags. Wrapped flowers produced seeds with no further manipulation, meaning that pollen transfer was intra-individual (self-pollination) only. That procedure excluded effects resulting from differences in mating system of the respective florets, which might contribute to ecological differences of the distinct seed types of heterocarpic plants (Olivieri et al. 1983; Cheptou et al. 2001). After 16 weeks of plant growth, shoots and seeds were harvested, while roots remained in the pots. Using seeds produced by plants of the first soil training, I started a second training round with 18 weeks of plant growth. Analogously to the seed morph scheme of the first soil training, I planted seedlings that originated from either non-pappus or papppus seeds, i.e. the identity of genotypes in the pots was maintained (Figure III.1). Again, plants of the second training round reproduced via self-pollination, which minimized genetic variability. As a result of the first and second soil training, I had non-pappus and pappus seeds produced by plants, which had either a non-pappus or pappus seed history for two plant generations and, further, corresponding soils trained by plants arising from either non-pappus or pappus seeds (Figure III.1). Trained soils and seeds produced by the last training were used for the final feedback experiment.

Before using non-pappus and pappus seeds in the final feedback step, I determined average seed weight of the different seed types produced per plant (in groups of 10 seeds each). During the feedback experiment, I separately collected all seeds developing from a single flower head (up to four capitula per plant if possible). From a total number of 731 capitula, I quantified numbers of non-pappus and pappus seeds produced. At harvest, I counted the number of capitula per plant and removed shoots from roots. Roots were separated from soil and carefully washed under a stream of water. Biomasses of roots and shoots were weighed after drying for five days at 40 °C. To assess mycorrhizal and other fungal root colonization, I used the ink-vinegar method of Vierheilig et al. (1998) and stained a representative root sample of each plant (ca. 100 mg of dry root material). Arbuscular mycorrhizal (AM) root colonization by structures of hyphae, arbuscules, vesicles and infection with non-AM fungi were determined using the magnified intersect method of McGonigle et al. (1990) based on 100 intersections per root sample examined at 200X.

In all experimental phases, plants grew in 6 x 25 cm, 400 ml Conetainer pots (Stuewe and Sons., Oregon, USA) in fully randomized arrangements. The first training step was completed from May until September 2010 under greenhouse conditions (natural photoperiod, day temperature 21-34 °C, night temperature 18-19 °C). The second training round and also the final feedback experiment were performed in a climate chamber (16 hour-photoperiod, day temperature 20 °C, night temperature 18 °C, humidity 60 %). Plants were always watered as needed with the same amount of water.

4.3.4 Statistical analyses

To calculate soil feedback responses I used the approach of 'home vs. away' contrasts according to other feedback studies (e.g. Bever 1994; Klironomos 2002). In my experiment, soil inoculated with trained soil represented 'home', and sterile soil was defined as 'away'. Soil feedback, therefore, was determined by subtracting the average measure of a given response variable in sterile soil from the average measure in soil inoculated with trained soil. Further, to calculate a value for individual investment in dispersal, I estimated the ratio of number of pappus seeds to total seeds per individuum (in contrast to Cheptou (2008) using number of non-dispersing seeds as the numerator).

To answer if plants from non-pappus and pappus seeds of *G. parviflora* differ in their soil feedback responses, I used linear mixed model analysis (Gotelli and Ellison 2004). The mixed model was specified with four factors as fixed effects (soil, seed history, seed type tested, soil treatment) and plant genotype as random effect. Treating genotype as random effect allows broader generalization of the results. Running the saturated model, I found no significant influences of the four-way and the three-way interactions on response variables tested (except for a significant interaction termed soil*seed type tested*soil treatment for total biomass). Therefore, I simplified the model to the hypothesized seed type-soil treatment interaction. Applying log-likelihood ratio tests, I tested if the model simplification was allowed, i.e. if the reduced model had a better fit than the saturated or respective higher-order interaction model.

To deal with heterogeneity within the data, I incorporated different variances per stratum in the model (Zuur et al. 2009) by using the 'varIdent' function in R (R Development Core Team 2009). Thus, the optimal model for number of total seeds and pappus seeds per plant was specified by a variance structure of varIdent(form=~1|soil). For root biomass and number of capitula per plant, models with varIdent(form=~1|soil*soil treatment) had the lowest AIC and, therefore, were selected. Total biomass data were analyzed with varIdent(form=~1|soil*soil history). Data estimated as dispersal ratio showed homogeneity, but were Box-Cox transformed (exponent 2) to meet assumptions of normality. To achieve normality, exponentiation was also needed for data of seed weight of seeds produced by the second training round (exponent −0.3434343). To normalize percentages of AM fungal root colonization by hyphae and arbuscules, I used arcsin square root transformation. Data of percent root colonization by non-AM fungi were log-transformed.

To identify differences in root colonization (AM fungal structures of hyphae, arbuscules, vesicles and non-AM fungi), I used the linear mixed model approach, where data from the treatment 'soil inoculated with trained soil' were analyzed exclusively (fixed effects: soil, seed history, seed type tested; random effect: plant genotype). Differences between treatment groups were always compared with posteriori pair-wise comparisons based on the resulting 95 % confidence limits (Zuur et al. 2009). For this purpose, standard errors (SE) of means and corresponding confidence limits were calculated from the mixed models that took the random effect of plant genotype into account. This procedure made the posteriori pair-wise comparisons very conservative.

Seed size of seeds produced during the second training round was analyzed by Analysis of Variance (three-way ANOVA; factors: soil, seed history and seed type; $P < 0.05$). All statistical analyses were performed using R version 2.10.1 (R Development Core Team 2009).

4.4 Results

Always starting with the removal of the highest order interaction effect, I found that the models without the four and all three-way interactions fitted the data best. Overall, reproductive output of *Galinsoga parviflora* was strongly influenced by soil treatment, i.e. if the soil was sterile (hereafter sterile) or inoculated with trained soil (hereafter trained) (Table III.1). In sterile soil, plants produced significantly more flower heads (mean ± SE calculated from the intercept of the mixed model; capitula sterile: 53.8 ± 1.7, trained: 42.0 ± 1.4), and had on average more seeds (total seeds sterile: 1668.1 ± 56.2, trained: 1469.6 ± 56.1). Numbers of total seeds and pappus seeds per plant were positively correlated (r = 0.996, Pearson's product-moment correlation).

4.4.1 Feedback contrasts of plants from the two seed types differ

The majority of response variables were significantly influenced by the interaction between soil treatment and seed type (non-pappus vs. pappus) (Table III.1). For plants arising from pappus seeds, root biomass, number of capitula, and numbers of total seeds and pappus seeds per plant were strongly increased in sterile soil compared to trained soil treatment (Table III.2). Plants grown from non-pappus seeds produced also significantly more capitula in sterile soil, but differences in root biomass, number of total seeds, as well as number of pappus seeds per plant were less pronounced. Therefore, soil feedback contrasts for root biomass, number of capitula, total seeds and pappus seeds per plant, were always larger for plants from pappus seeds than from non-pappus seeds (Figures III.2a–c). Dispersal ratio for plants arising from non-pappus seeds showed a significant shift towards a higher portion of pappus seeds in trained soil treatment compared to all other combinations of soil treatment and seed type tested (Table III.2). Calculated soil feedback for dispersal ratio of plants arising from non-pappus seeds was positive, while tended to be negative for plants from pappus seeds (Figure III.2d).

Seed type–soil treatment patterns, however, had no correspondence with root colonization by pathogenic fungi. In the trained soil treatment, roots were extremely rare colonized by non-AM fungi (trained: 0.06 ± 0.04 %). I detected no significant effect of soil, seed history or seed type tested on non-AM fungal root colonization (Supplemental Table A.III.1). Furthermore, I found no significant effect of any of the factors tested on root colonization by AM hyphae and AM vesicles (Supplemental Table A.III.1). AM hyphal root colonization was similar in both inoculated soils (trained Dahlem: 81.6 ± 3.8 %, trained Thyrow: 80.1 ± 3.8 %).

However, colonization by AM arbuscules in trained soil treatment indicated a strong soil effect (soil: $P < 0.001$, mixed effect model). When plants grew in soil inoculated with trained soil from Dahlem, percentage root colonization by arbuscules was significantly higher than for plants growing in soil treated with trained soil from Thyrow (root colonization by arbuscules in trained soil Dahlem: 65.4 ± 3.9 %, trained soil Thyrow: 38.5 ± 3.9 %).

4.4.2 Effects of soil and the soil–soil treatment interaction

Depending on the soil type, biomass measures and dispersal ratio differed significantly (Table III.1). Overall, plants growing in Thyrow soil had greater root biomass (root biomass in Thyrow soil: 0.444 ± 0.032 g, Dahlem soil: 0.323 ± 0.034 g), as well as more total biomass (total biomass in Thyrow soil: 1.743 ± 0.132 g, Dahlem soil: 1.295 ± 0.141 g). The dispersal ratio was also strongly increased for plants in Thyrow soil (dispersal ratio in Thyrow soil: 0.775 ± 0.008, Dahlem soil: 0.732 ± 0.008).

A significant interaction between soil and soil treatment was found for total biomass, root biomass, and number of capitula per plant (Table III.1). In Dahlem soil, the presence of trained soil inoculum had a strong growth reducing effect (Table III.3). Further, number of capitula per plant was significantly increased in sterile Dahlem soil compared to all other combinations of soil and treatment tested (Table III.3). In Thyrow soil, however, inoculation with trained soil decreased number of capitula per plant, while root and total biomass were minimally influenced compared to the sterile treatment.

4.4.3 Effects of the seed history–seed type interaction

Plant root biomass was significantly influenced by the interaction between seed type and seed history over two plant generations (Table III.2). Plants grown from pappus seeds (papp), which also had a pappus history (papp hist), produced the smallest root biomass (root biomass for papp with papp hist: 0.346 ± 0.039 g). However, plants grown from pappus seeds, which had a non-pappus history (non-papp hist), had the greatest root biomass compared to all other combination of seed history and seed type tested (root biomass for papp with non-papp hist: 0.435 ± 0.039 g). For plants from non-pappus seeds (non-papp), the influence of seed history on root biomass tended in the same direction, but was less pronounced (root biomass for non-papp with papp hist: 0.368 ± 0.039 g, non-papp with non-papp hist: 0.403 ± 0.039 g). Overall, root biomass of plants grown from the distinct seed types differed marginally; the indicated significant seed type effect could not be shown with the 95 % confidence limit (root biomass non-papp: 0.386 ± 0.028 g, papp: 0.391 ± 0.028 g).

Further, seed weight of seeds produced by plants of the second training round indicated a strong seed type effect (Supplemental Table A.III.2, seed type: $F_{1,23}=38.27$; $P < 0.001$, ANOVA). Non-pappus seeds were significantly heavier than pappus seeds (seed weight non-papp: 167.9 ± 2.7 µg, papp: 145.0 ± 2.9 µg).

4.5 Discussion

Under climate chamber conditions, plants arising from the distinct seed morphs of *G. parviflora* differed in their soil feedback responses. The soil feedback differences between progeny of non-pappus and pappus seeds were found for reproductive traits, such as number of capitula, number of total seeds, as well as number of pappus seeds per plant, but were also indicated by root biomass. Consistent with the hypothesis, progeny of non-pappus seeds, which are produced for *in situ* persistence, were less negatively affected by conditions of trained ('self-cultivated') soil compared to progeny of pappus seeds, which easily disperse from the capitula (Espinosa-García et al. 2003) and represent the long-distance dispersal type (Vibrans 1999). Therefore, the magnitude of negative soil feedback corresponded to the dispersal potential of the two seed types giving an advantage to plants from non-pappus seeds in the existing population.

For dispersal ratio, I even found positive feedback for non-pappus progeny. They proportionally produced more seeds equipped with a pappus when the soil was inoculated with trained soil. Thus, non-pappus progeny showed greater investment in dispersal under unfavorable conditions of trained soil. This strategy would allow escape from sib competition and negative density effects (Venable and Brown 1988), although pappus seeds of *G. parviflora* have a higher risk of failure than non-pappus seeds (Venable 1985a, Simons 2011). Pappus progeny, in contrast, did not alter the ratio of number of pappus seeds to total seeds per individuum depending on the soil treatment; hence, showed neutral soil feedback for dispersal ratio. Therefore, adverse environmental conditions led to a proportional shift between non-pappus and pappus seeds in non-pappus progeny only. To my knowledge, this is the first evidence that shifts in the ratio of particular seed types produced by a heterocarpic plant species are associated with the seed morph from which the plant was arising. Besides, the result confirms other observations that dispersal ratio reflects a highly variable trait by which heterocarpic species adjust to environmental conditions (Mandák 1997). For example, nutrient depletion in *Crepis sancta* (Imbert and Ronce 2001) or high competition in *Hypochoeris glabra* (Baker and O'Dowd 1982) increased proportion of dispersing seeds, while habitat fragmentation in *C. sancta* (Cheptou et al. 2008) and increasing aridity in *Hedypnois rhagadioles* (Kigel 1992) or the genus *Picris* along a gradient from mesic to arid (Ellner and Schmida 1984) led to a higher proportion of non-dispersing seeds.

Interestingly, divergent soil feedback responses of plants grown from the different seed types of *G. parviflora* did not coincide with differences in fungal root colonization. Overall, percentage of root infection with pathogenic fungi in the trained soil treatment was very low and not impacted by factors tested. Thus, I found no evidence for non-pappus progeny being less colonized and, therefore, less negatively affected by fungal pathogens. Furthermore, I did not find significant influences of the seed type tested on root colonization by AM fungi, which can also generate negative soil feedback (Bever 2002). Consequently, differences in soil feedback of non-pappus and pappus progeny must have been caused by other microorganisms of the very complex soil community (Bever 2003): possibly nematode pathogens (De Deyn et al. 2003) or bacterial antagonists, which I did not evaluate in this study. However, influences of soil bacteria and fungi on the different seed morphs of *G. parviflora* were investigated by Espinosa-García et al. (2003). Focusing on differential longevity of the two seed types in soil, they found no correlation between different loss rates from the seed bank and their susceptibility to fungi or bacteria. Further, they report low incidence of fungal infection in the seeds. Given the fact that I also rarely observed colonization by non-AM fungi in the plant roots, pathogenic fungi may be of minor importance for *G. parviflora*.

However, in contrast to Espinosa-García et al. (2003) studying *G. parviflora* in the native region (Damalas 2008), my experiment was conducted in the introduced range, where most likely generalist pathogens rather than species-specific play a role (Kulmatiski et al. 2008). As Klironomos (2002) demonstrated, introduced species often accumulate pathogens slowly and, therefore, show neutral feedback in the presence of a 'self-cultivated' pathogen/saprobe fraction only. But, since I found negative soil feedback using whole soil as inoculum, the performed soil training over two plant generations was sufficient to generate an adequate pathogen load in the soil. Nevertheless, it would be interesting to test soil feedback responses of plants arising from the two seed types to different soil fractions, and to disentangle what kind of soil biota might have caused the greater negative soil feedback for plants from pappus compared to non-pappus seeds.

Further, my data show that plants grown from pappus seeds had significantly more capitula and greater reproductive output under sterile soil conditions than those from non-pappus seeds. In sterile soil, therefore, the tendency of a colonizer to produce large numbers of viable seeds (Warwick and Sweet 1983) was more pronounced in progeny of pappus than non-pappus seeds. Thus, greater dispersal capacity of pappus seeds in *G. parviflora* might be associated with higher investment in reproductive output. Such a relationship would be adaptive because it allows pappus progeny a rapid population buildup after pappus seeds have reached new habitats.

Conversely, I did not find that non-pappus progeny, which maintain the existing population, invested more in biomass. This was surprising given the greater seed size of non-pappus compared to pappus seeds, and the observed positive relationship between seed weight and seedling growth after 9 days of emergence (Rai and Tripathi 1982). However, the positive correlation between seed size and plant growth/reproductive output was also shown to depend on environmental conditions, such as competition and nutrient deficiency. Rai and Tripathi (1987) planted seedlings arising from the different seed types in monoculture and mixture, and demonstrated that greater seed weight of non-pappus seeds is advantageous at low or medium fertilizer doses only. At high fertilizer dose, plants from smaller pappus seeds performed better; hence, soil nutrient status altered the competitive ability of non-pappus and pappus progeny. Other studies of heterocarpic Asteraceae also report that greater seed size is not always converted into biomass growth, but becomes advantageous in the presence of competition (Imbert et al. 1997; De Clavijo and Jiménez 1998).

Further, the results indicate that the growth of plants arising from the two seed types was influenced by the seed type of the mother/grandmother plant. In the final feedback experiment, pappus progeny produced significantly more root biomass when they had a non-pappus history, i.e. came from non-pappus mother/grandmother plants. For non-pappus progeny, I did not find such a seed history influence. Thus, by having a non-pappus mother/grandmother pappus progeny showed increased belowground growth, which might be advantageous in terms of improved efficiency of plant nutrient uptake (Imbert et al. 1997) and occupance of space in a new colonized patch, respectively. Therefore, further research into the phenomenon of heterocarpic species should not only consider the seed type of a plant, but also the seed type of prior plant generations.

To conclude, plants arising from the two distinct seed morphs of *G. parviflora* differ in their soil feedback responses: plants from non-pappus seeds are less affected by unfavorable conditions of trained ('self-cultivated') soil than plants from pappus seeds. Beside soil feedback, the differentiation between the two seed morphs in *G. parviflora* involves dispersal potential, dispersal pathway (Vibrans 1999), viability in the soil (Espinosa-García et al. 2003), as well as competitive ability (Rai and Tripathi 1987). Therefore, the strategies of the two seed types are very complex: greater dispersal ability away from maternal plants of lighter pappus seeds is accompanied with shorter survival in the seed bank, and reduced performance of plants arising from pappus seeds in home soil, as well as lower competitive ability in mixed populations under moderate soil fertility compared to plants from non-pappus seeds, which experience less negative soil feedback. Therefore, in mixed populations of plants arising from non-pappus and pappus seeds less negative soil feedback and greater competitive ability of plants from non-pappus seeds give an advantage to non-pappus progeny in the local habitat.

4.6 References

Aguado M, Martínez-Sánchez JJ, Reig-Armiñana J, García-Breijo FJ, Franco JA, Vicente MJ (2011) Morphology, anatomy and germination response of heteromorphic achenes of *Anthemis chrysantha* J. Gay (Asteraceae), a critically endangered species. Seed Sci Res 21:283-294.

Baker GA, O'Dowd DJ (1982) Effects of parent plant density on the production of achene types in the annual *Hypochoeris glabra*. J Ecol 70:201-215.

Baskin JM, Baskin CC (1976) Germination dimorphism in *Heterotheca subaxillaris* var. *subaxillaris*. B Torrey Bot Club 103:201-206.

Baumecker M, Ellmer F, Köhn W (2009) Statischer Düngungs- und Beregnungsversuch Thyrow. In: MLUV & LVLF Brandenburg (ed.), Dauerfeldversuche Brandenburg und Berlin. Beiträge für eine nachhaltige landwirtschaftliche Bodenbenutzung, Schriftenreihe des Landesamtes für Verbraucherschutz, Landwirtschaft und Flurneuordnung, Abteilung Landwirtschaft und Gartenbau, pp. 142-147.

Becker W (1913) Über die Keimung verschiedenartiger Früchte und Samen bei derselben Spezies. Beihefte Botanisches Centralblatt 29:21-143.

Bever JD (1994) Feedback between plants and their soil communities in an old field community. Ecology 75:1965-1977.

Bever JD (2002) Negative feedback within a mutualism: host-specific growth of mycorrhizal fungi reduces plant benefit. Proc R Soc Lond B 269:2595-2601.

Bever JD (2003) Soil community feedback and the coexistence of competitors: conceptual frameworks and empirical tests. New Phytol 157:465-473.

Brändel M (2004) Dormancy and germination of heteromorphic achenes of *Bidens frondosa*. Flora 199:228-233.

Brändel M (2007) Ecology of Achene Dimorphism in *Leontodon saxatilis*. Ann Bot-London 100:1189-1197.

Canne JM (1977) A revision of the genus *Galinsoga* (Compositae: Helianthae). Rhodora 79:319-389.

Cheptou PO, Carrue O, Rouifed S, Cantarel A (2008) Rapid evolution of seed dispersal in an urban environment in the weed *Crepis sancta*. PNAS 105:3796-3799.

Cheptou PO, Lepart J, Escarre J (2001) Differential outcrossing rates in dispersing and non-dispersing achenes in the heterocarpic plant *Crepis sancta* (Asteraceae). Evol Ecol 15:1-13.

Cruz-Mazo G, Buide ML, Samuel R, Narbona E (2009) Molecular phylogeny of *Scorzoneroides* (Asteraceae): Evolution of heterocarpy and annual habit in unpredictable environments. Mol Phylogenet and Evol 53:835-847.

Damalas CA (2008) Distribution, biology, and agricultural importance of *Galinsoga parviflora* (Asteraceae). Weed Biol Manag 8:147-153.

De Clavijo ER (2001) The role of dimorphic achenes in the biology of the annual weed *Leontodon longirrostris*. Weed Res 41:275-286.

De Clavijo ER, Jiménez MJ (1998) The influence of achene type and plant density on growth and biomass allocation in the heterocarpic annual *Catananche lutea* (Asteraceae). Int J Plant Sci 159:637-647.

De Deyn GB, Raaijmakers CE, Zoomer HR, Berg MP, de Ruiter PC, Verhoef HA, Bezemer TM, van der Putten WH (2003) Soil invertebrate fauna enhances grassland succession and diversity. Nature 442:711-713.

Ehrenfeld JG, Ravit B, Elgersma K (2005) Feedback in the plant–soil system. Annu Rev Environ Resour 30:75-115.

Ellison AM (1987) Effect of seed dimorphism on the density-dependent dynamics of experimental populations of *Atriplex triangularis* (Chenopodiaceae). Am J Bot 74:1280-1288.

Ellmer F, Peschke H, Köhn W, Chmielewski FM, Baumecker M (2000) Tillage and fertilizing effects on sandy soils. Review and selected results of long-term experiments at Humboldt-University Berlin. J Plant Nutr Soil Sci 163:267-272.

Ellner SP, Schmida A (1984). Seed dispersal in relation to habitat in the genus *Picris* (Compositae) in Mediterranean and arid regions. Isr J Bot 33:25-39.

Espinosa-García FJ, Sarukhán J (1997) Manual de malezas del Valle de México: claves, descripciones e ilustraciones, Universidad Nacional Autónoma de México, México, p. 84.

Espinosa-García FJ, Vázquez-Bravo R, Martínez-Ramos M (2003) Survival, germinability and fungal colonization of dimorphic achenes of the annual weed *Galinsoga parviflora* buried in the soil. Weed Res 43:269-275.

Flint SD, Palmblad IG (1978) Germination dimorphism and developmental flexibility in the ruderal weed *Heterotheca grandiflora*. Oecologia 36:33-43.

Gotelli NJ, Ellison AM (2004) A Primer of Ecological Statistics, Sinauer Associates, Sunderland, MA.

Harper JL, Lovell PH, Moore KG (1970) The shapes and sizes of seeds. Annu Rev Ecol Syst 1:327-356.

Holm LG, Plucknett DL, Pancho JV, Herberger JP (1977) The world's worst weeds: distribution and biology, University Press of Hawaii, Honolulu.

Imbert E (2002) Ecological consequences and ontogeny of seed heteromorphism. Perspect Plant Ecol 5:13-36.

Imbert E, Escarré J, Lepart J (1996) Achene dimorphism and among-population variation in *Crepis sancta* (Asteraceae). Int J Plant Sci 157:309-315.

Imbert E, Escarré J, Lepart J (1997) Seed heteromorphism in *Crepis sancta* (Asteraceae): performance of two morphs in different environments. Oikos 79:325-332.

Imbert E, Ronce O (2001) Phenotypic plasticity for dispersal ability in the seed heteromorphic *Crepis sancta* (Asteraceae). Oikos 93:126-134.

Johnson DJ, Beaulieu WT, Bever JD, Clay K (2012) Conspecific negative density dependence and forest diversity. Science 336:904-907.

Kigel J (1992) Diaspore heteromorphism and germination in populations of the ephemeral *Hedypnois rhagadioloides* (L.) F.W. Schmidt (Asteraceae) inhabiting a geographic range of increasing aridity. Acta Oecol 13:45-53.

Klironomos JN (2002) Feedback with soil biota contributes to plant rarity and invasiveness in communities. Nature 417:67-70.

Kulmatiski A, Beard KH, Stevens JR, Cobbold SM (2008) Plant-soil feedback: a meta-analytical review. Ecol Lett 11:980-992.

Mandák B (1997) Seed heteromorphism and plant life cycle: a review of literature. Preslia 69:129-159.

Mandák B, Pyšek P (2005) How does seed heteromorphism influence the life history stages of *Atriplex sagittata* (Chenopodiaceae)? Flora 200:516-526.

McEvoy PB (1984) Dormancy and dispersal in dimorphic achenes of tansy ragwort, *Senecio jacobaea* L. (Compositae). Oecologia 61:160-168.

McGonigle TP, Miller MH, Evans DG, Fairchild GL, Swan JA (1990) A new method which gives an objective measure of colonization of roots by vesicular arbuscular mycorrhizal fungi. New Phytol 115:495-501.

Nielreich A (1866) Nachträge zur Flora Nieder-Oesterreich, K. k. zoologisch-botanische Gesellschaft in Wien, F. A. Brookhaus, Leipzig.

Olivieri I, Swan M, Gouyon PH (1983) Reproductive system and colonizing strategy of two species of *Carduus* (Compositae). Oecologia 60:114-117.

Petermann JS, Fergus AJF, Turnbull LA, Schmid B (2008) Janzen-Connell effects are widespread and strong enough to maintain diversity in grasslands. Ecology 89:2399-2406.

Philippi T, Seger J (1989) Hedging one's evolutionary bets, revisited. Trends Ecol Evol 4:41-44.

R Development Core Team (2009) R: A Language and Environment for Statistical Computing. R Foundation for Statistical Computing, Vienna, Austria. http://www.R-project.org.

Rai JPN, Tripathi RS (1982) Adaptive significance of seed reserves in ray achenes of *Galinsoga parviflora* Cav. Experientia 38:804-805.

Rai JPN, Tripathi RS (1987) Germination and plant survival and growth of *Galinsoga parviflora* Cav. as related to food and energy content of its ray- and disc-achenes. Acta Oecol 8:155-165.

Rocha OJ (1996) The effects of achene heteromorphism on the dispersal capacity of *Bidens pilosa* L. Int J Plant Sci 157:316-322.

Schweitzer K (2010) Naturnahe Böden im Berliner Stadtgebiet. Referenz für anthropogene Bodenentwicklung oder Beispiel für anthropogene Bodenveränderungen. In: Makki M, Frielinghaus M (eds) Boden des Jahres 2010 - Stadtböden. Berlin und seine Böden. Berliner Geographische Arbeiten, Humboldt-Universität, Berlin, pp. 13-22.

Simons AM (2011) Modes of response to environmental change and the elusive empirical evidence for bet hedging. Proc R Soc B 278:1601-1609.

Sorensen AE (1978) Somatic polymorphism and seed dispersal. Nature 276:174-176.

Sun HZ, Lu JJ, Tan DY, Baskin JM, Baskin CC (2009) Dormancy and germination characteristics of the trimorphic achenes of *Garhadiolus papposus* (Asteraceae), an annual ephemeral from the Junggar Desert, China. S Afr J Bot 75:537-545.

Tanowitz BD, Salopek PF, Mahall BE (1987) Differential germination of ray and disc achenes in *Hemizonia increscens* (Asteraceae). Am J Bot 74:303-312.

Terzioğlu S, Anşin R (2001) A chorological study of the taxa naturalized in the Eastern Black Sea Region. Turk J Agric For 25:305-309.

Venable DL (1985a) The evolutionary ecology of seed heteromorphism. Am Nat 126:577-595.

Venable DL (1989) Modeling the evolutionary ecology of seed banks. In: Leck MA, Parker VT, Simpson RL (eds) The ecology of soil seed banks, Academic Press, San Diego, pp. 67-87.

Venable DL (2007) Bet hedging in a guild of desert annuals. Ecology 88:1086-1090.

Venable DL, Brown JS (1988) The selective interactions of dispersal, dormancy, and seed size as adaptations for reducing risk in variable environments. Am Nat 131: 360-384.

Venable DL, Dyreson E, Morales E (1995) Population dynamic consequences and evolution of seed traits of *Heterosperma pinnatum* (Asteraceae). Am J Bot 82:410-420.

Venable DL, Levin DA (1985a) Ecology of achene dimorphism in *Heterotheca latifolia* II. Demographic variation within populations. J Ecol 73:743-755.

Venable DL, Levin DA (1985b) Ecology of achene dimorphism in *Heterotheca latifolia* I. Achene structure, germination and dispersal. J Ecol 73:133-145.

Venable DL (1985b) Ecology of achene dimorphismin *Heterotheca latifolia*. III. Consequences of varied water availability. J Ecol 73:757-763.

Vibrans H (1999) Epianthropochory in Mexican weed communities. Am J Bot 86:476-481.

Vierheilig H, Coughlan AP, Wyss U, Piche Y (1998) Ink and vinegar, a simple staining technique for arbuscular-mycorrhizal fungi. Appl Environ Microb 12:5004-5007.

Warwick SI, Sweet RD (1983) The biology of Canadian weeds. 58. *Galinsoga parviflora* and *G. quadriradiata* (= *G. ciliata*). Can J Plant Sci 63:695-709.

Zuur AF, Ieno EN, Walker NJ, Saveliev AA, Smith GM (2009) Mixed effect models and extensions in ecology with R, Springer, NY, USA.

4.7 Tables and figures

Table III.1 Mixed effect model analysis on responses of biomass and reproductive traits of *Galinsoga parviflora* in the feedback experiment, with soil, seed history (Hist), seed type, and soil treatment (Treat) as factors. Plant genotype was treated as a random effect. Values in bold indicate significance at $P < 0.05$.

Factors	Total biomass P	Root biomass P	No. capitula P	Total seeds P	Papp. seeds P	Dispersal ratio P
Soil	**0.0441**	**0.018**	0.406	0.504	0.316	**0.003**
Hist	0.3449	0.292	0.861	0.219	0.133	0.609
Seed.type	0.1132	**0.050**	0.447	0.114	0.079	0.213
Treat	0.1527	0.191	**<0.001**	**0.001**	**0.001**	0.120
Soil x Hist	0.954	0.987	0.961	0.333	0.261	0.072
Soil x Seed.type	0.960	0.690	0.675	0.617	0.610	0.659
Hist x Seed.type	0.090	**0.020**	0.653	0.873	0.996	0.345
Soil x Treat	**0.001**	**0.001**	**0.018**	0.077	0.102	0.681
Hist x Treat	0.927	0.162	0.391	0.119	0.111	0.282
Seed.type x Treat	0.581	**0.049**	**0.034**	**0.037**	**0.017**	**0.003**

No. capitula, number of capitula per plant; Total seeds, average number of total seeds per plant; Papp. seeds, average number of pappus seeds per plant; Dispersal ratio, ratio of number of pappus seeds to total seeds per individuum.

Table III.2 Response variables of plants from the distinct seed types of *Galinsoga parviflora* to soil treatment in the experiment. Values of mean ± SE were calculated from the intercept of the mixed effect model; hence, plant genotype was always incorporated in SE. Different lower case letters indicate differences according to 95 % confidence limit.

	Plants from non-pappus seeds		Plants from pappus seeds	
	sterile soil	trained soil	sterile soil	trained soil
Root bio. (g)	0.396 ± 0.031 [ab]	0.370 ± 0.030 [b]	0.433 ± 0.032 [a]	0.360 ± 0.030 [b]
No. capitula	51.7 ± 2.2 [b]	43.1 ± 1.6 [c]	56.0 ± 2.2 [a]	40.8 ± 1.6 [c]
Papp. seeds	1370.5 ± 65.9 [b]	1338.3 ± 65.9 [bc]	1539.1 ± 66.2 [a]	1230.3 ± 65.9 [c]
Total seeds	1581.5 ± 71.5 [b]	1514.1 ± 71.5 [bc]	1755.7 ± 71.8 [a]	1424.1 ± 71.5 [c]
Disp. ratio	0.742 ± 0.01 [b]	0.771 ± 0.01 [a]	0.756 ± 0.01 [ab]	0.745 ± 0.01 [b]

Root bio., root biomass; No. capitula, number of capitula per plant; Papp. seeds, number of pappus seeds per plant; Disp. ratio, dispersal ratio, i.e. ratio of number of pappus seeds to total seeds per individuum.

Table III.3 Response variables of plants from the distinct seed types of *Galinsoga parviflora* to soil and soil treatment. Values of mean ± SE were calculated from the intercept of the mixed effect model with genotype as a random factor. Different lower case letters indicate differences according to 95 % confidence limit.

	Dahlem soil		Thyrow soil	
	sterile	trained soil	sterile	trained soil
Total bio. (g)	1.498 ± 0.148 [a]	1.098 ± 0.147 [b]	1.785 ± 0.134 [a]	1.700 ± 0.134 [a]
Root bio. (g)	0.369 ± 0.035 [b]	0.252 ± 0.036 [c]	0.463 ± 0.034 [a]	0.438 ± 0.032 [a]
No. capitula	58.0 ± 3.4 [a]	37.5 ± 2.9 [c]	53.1 ± 2.1 [b]	42.8 ± 1.7 [c]

Total bio., total biomass; Root bio., root biomass; No. capitula, number of capitula per plant.

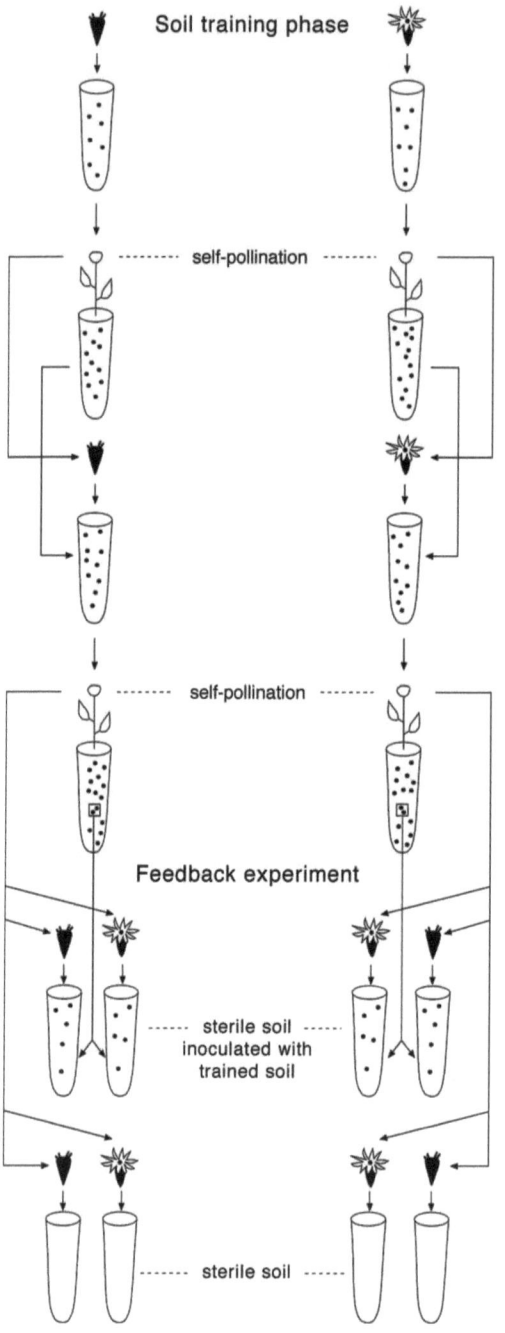

Figure III.1 Design of the feedback experiment with a soil training phase performed over two plant generations. During the first training step, the initial soil was trained by plants arising from either non-pappus or pappus seeds of *Galinsoga parviflora*. For the second training round, roots of plants of the first training remained in the pots, and plants (first training progeny) were analogously planted to the first training scheme. In all training phases, plants reproduced by self-fertilization, where flowering stems with unopened capitula were separately enclosed in paper bags allowing intra-individual pollen transfer (self-pollination) only. For the feedback step, progeny and soil of the second training round were used. To prepare 'soil inoculated with trained soil', trained soil was thoroughly mixed into sterile background soil (ratio 1:10). Identity of replicates was always maintained.

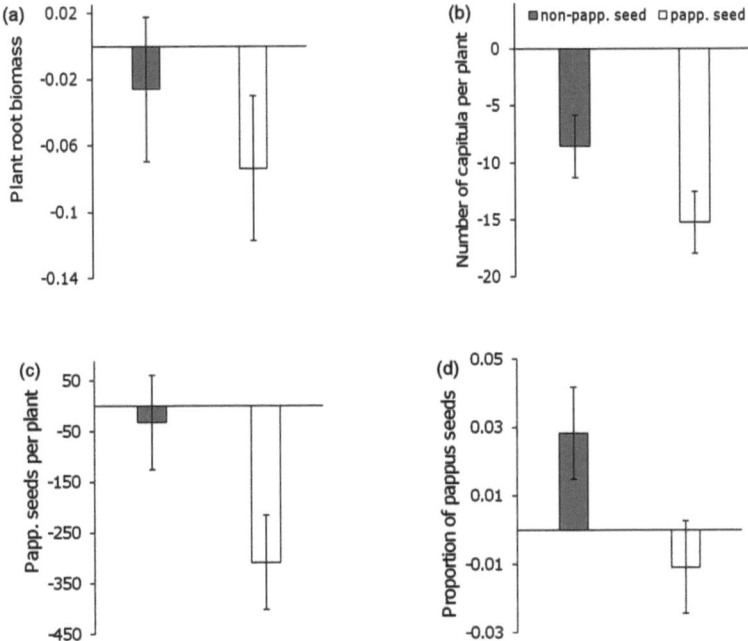

Figure III.2 'Trained vs. sterile' soil contrasts of plants grown from non-pappus (grey) and pappus seeds (white) of *Galinsoga parviflora* in the experiment: (a) plant root biomass; (b) number of capitula per plant; (c) pappus seeds per plant; (d) dispersal ratio. Contrasts were determined by subtracting the average measure of a given response variable in sterile soil from the average measure in soil inoculated with trained soil, which follows the calculation of 'home vs. away' contrasts in plant–soil feedback studies (Bever 1994; Klironomos 2002). Error bars were calculated from the errors (SE) of the linear mixed effect model with genotype as a random factor.

5 SUMMARY

Numerous ecosystems worldwide are influenced by invasive species in their functioning in a multitude of ways (Wardle et al. 2011). Invasive species, moreover, can damage ecosystem services that are fundamental to human well-being resulting in substantial economic costs (e.g. Pimentel et al. 2005; Pejchar and Mooney 2009; Vilà et al. 2010). Therefore, it is of increasing urgency to better understand the mechanisms involved in the invasion process as human activities such as international trade, transport and travel, which cause species dispersal into new ranges, continue to expand (Keller et al. 2011).

The success of invasive plant species must be regarded as highly context-dependent and linked to a combination of both abiotic and biotic factors, and multiple mechanisms (e.g. Daehler 2003; Richardson and Pyšek 2006; Barney and Whitlow 2008). In particular, biotic interactions of invasive species with their new environment may be the key driver for the successful spread into new areas (Jeschke et al. 2012). However, despite a large number of articles published both on experimental and theoretical topics in invasion biology per year (Kühn et al. 2011), it is still difficult to give precise statements on why a particular plant species becomes a dominant component in a plant community where it is not native.

In the book, I report about the interaction of invasive plant species with belowground organisms, especially arbuscular mycorrhizal (AM) fungi. I conducted a series of experiments to examine the importance of the AM fungal association for the successful spread of *Ambrosia artemisiifolia* in the new European range; *A. artemisiifolia* has been proposed to be facilitated by the symbiosis with AM fungi in Central Europe (Fumanal et al. 2006). Further, I focused on the phenomenon of heterocarpy of the non-native plant *Galinsoga parviflora* in a soil feedback study. I investigated if the different dispersal capacities of the two distinct seed types correlate to different soil feedback responses, which may contribute to the success of

G. parviflora in the new range. In all experiments, I always maintained the ecological context of soil and natural AM fungal community; hence, my findings have a comparatively high realism. Below I summarize the main results from the three studies.

Study I: Divergent responses of *Ambrosia artemisiifolia* to natural AM fungal communities in the new European range.

Background and Aims: AM fungi may act more or less cooperatively in association with plants exhibiting functions from mutualism to parasitism depending on soil and light environments (Johnson et al. 1997; Kiers et al. 2011). Therefore, natural selection pressure in arbuscular mycorrhizas may favor the AM fungi–plant combinations that are the most fit under their respective local environmental circumstances, promoting local adaptation and co-adaptation in AM associations (Helgason and Fitter 2009; Hoeksema 2010). Recently, existence of coadapted AM fungal–plant interactions has been found (Johnson et al. 2010; Ji et al.2010), but knowledge about the extent to which such adaptations also occur during plant invasions is lacking. In this study, I investigated whether or not the mycorrhizal symbiosis between *A. artemisiifolia* and native AM fungal communities shows evidence of co-adaptation in the new European range. In a 'local vs. foreign' reciprocal inoculation experiment, I compared performance of plant genotypes from two different sites: a roadside and a cornfield habitat.

Results: Natural AM fungal assemblages were found to be mutualistic with *A. artemisiifolia* in low fertility roadside soil, but not in agricultural soil (Figure I.3a). Decreased plant growth in response to the less cooperative quality of the agricultural AM fungal community in the agricultural system coincided with alterations of plant root systems towards greater fineness. I found no evidence for locally adapted plant-AM fungal interactions, but adaptation of roadside plants to a 'local' roadside soil environment (Figure I.4). Further, soil conditions had a strong effect: plants growing in more fertile cornfield soil produced more biomass, had greater total seed weight, and flowered earlier than plants in less fertile roadside soil.

Conclusion: The results of this study indicate that performance of non-native *A. artemisiifolia* may be influenced by different mycorrhizal functions, leading to mutualism in less fertile roadside habitats and parasitism in more fertile cornfield soils. Moreover, the study highlights the great importance of the soil context for plant responses to mycorrhizal inoculation, and that mycorrhizal functions may be unpredictable when AM fungal communities are introduced to novel soils. Further, the findings show that adaptation to soil conditions may play a crucial role in the early stages of the spread of *A. artemisiifolia* along the road.

Study II: Non-native *Ambrosia artemisiifolia* are more influenced by relative density and identity of neighboring plant species than arbuscular mycorrhiza.

Background and Aims: One way by which invasive plants can interact with AM fungi has been described as the enhanced mutualisms hypothesis or facilitation, whereby invasive plants are positively influenced by the AM fungal association of the new range to the detriment of native species (Reinhart and Callaway 2006; Shah et al. 2009). In this context, several studies have shown that AM fungi may contribute to the dominance of invasive over native plants by altering competitive interactions (e.g. Marler et al. 1999; Callaway et al. 2004b; Shah et al. 2008). In this study, I investigated the effects of natural AM fungal communities on the competitive ability of *A. artemisiifolia* in two greenhouse experiments always maintaining the ecological context of soil and AM fungi. I studied *A. artemisiifolia* grown together with one of four co-existing mycorrhizal plant species in a 1:4 (target) or 4:1 (challenger) relative density. As neighbor species I selected *Conyza canadensis*, *Artemisia vulgaris*, *Daucus carota* and *Tanacetum vulgare*, which I found co-occurring with *A. artemisiifolia* in ruderal communities. Moreover, I studied the influence of a roadside AM fungal community on *A. artemisiifolia* and *D. carota* grown in pairwise situations of intra- and interspecific competition.

Results: Regardless of presence/absence of AM fungal communities, *A. artemisiifolia* was highly dominant in all interspecific competitive arrangements under the nutrient poor soil conditions tested. Divergent AM fungal effects on biomass of *A. artemisiifolia* as a function of neighbor plant were only indicated as trends: target *A. artemisiifolia* tended to either increase in shoot biomass (with *C. canadensis* or *T. vulgare*), decrease (with *D. carota*) or was unaffected (with *A. vulgaris*) under presence of mycorrhiza (Figure II.3a). In pairwise competitive situations, roadside AM fungi had an amplifying (negative) effect on *A. artemisiifolia* in intraspecific competition, and a neutral effect in mixture with *D. carota*. Moreover, *A. artemisiifolia* experienced strong competition by conspecifics, which caused decreases in shoot biomass and number of male inflorescences, but earlier flowering of female flowers. Among the mycorrhizal plant competitors, *C. canadensis* performed poorly compared to the other neighboring species tested.

Conclusion: The results of the experiments demonstrate that *A. artemisiifolia* is an exceptionally good competitor both at low and high relative density in comparison to co-existing mycorrhizal plant species in the new European range under nutrient poor soil conditions. Moreover, my findings suggest that the competitive ability of *A. artemisiifolia* – a successful pioneer plant and a species with a strongly ruderal life history – is very weakly influenced by natural AM fungal communities in the presence of other mycorrhizal plants in low fertility soils. Therefore, the invasive success of *A. artemisiifolia* in Central Europe may not be related to facilitation by natural AM fungal communities.

Study III: Distinct seed morphs of *Galinsoga parviflora* (Asteraceae) give rise to different soil feedbacks.

Background and Aims Heterocarpy is the phenomenon that a single plant produces two or more distinct fruit types, which often differ in dispersal mechanisms and ecological behavior. Beside fruit, the terms diaspore or seed are also used: I refer to seed *sensu lato*. Heterocarpy has been extensively studied from the perspective of seed size, dormancy and germination behavior (Mandák 1997). Some studies, further, report about differences in competitive ability of plants arising from different seed morphs (e.g. Rai and Tripathi 1987; Imbert et al. 1997) or the influence of environmental conditions (urban habitat fragmentation) on the ratio of non-dispersing to dispersing seeds of a heterocarpic plant species (Cheptou et al. 2008). In this study, I investigated another aspect of ecological behavior. I asked if plants arising from distinct seed types differ in their interaction with soil biota, i.e., exhibit divergent responses of soil feedback. I studied the non-native plant *G. parviflora*, which produces seeds both equipped with a pappus for long-distance dispersal and seeds without a pappus (non-pappus) for maintaining the existing population. The hypothesis was that plants arising from non-pappus seeds would exhibit better performance, i.e., less negative soil feedback, in soil trained by the mother plant than plants grown from pappus seeds. To test this, I trained soil over two plant generations with plants either arising from non-pappus or pappus seeds, and studied feedback responses of pappus and non-pappus progeny as 'trained versus sterile' soil contrasts (Figure III.1).

Results: Progeny grown from distinct seed morphs of *G. parviflora* differed in soil feedback. Plants grown from pappus seeds had greater root biomass and produced more flower heads, as well as total number of seeds and pappus seeds per plant in sterile compared to 'self-cultivated' (trained) soil conditions. For plants arising from non-pappus seeds the differences between sterile and trained soil treatment were less pronounced. Hence, negative feedback contrasts for the above mentioned response variables were always larger for plants arising from pappus seeds than those from non-pappus seeds (Figures II.2a–c). Progeny of non-pappus seeds, moreover, showed a significant shift towards a higher portion of pappus seeds in trained soil compared to all other combinations of soil treatment and seed type tested (Figure II.2d). Further, the data indicated that plants from pappus seeds produced significantly more root biomass when they had a non-pappus history, i.e. came from non-pappus mother/grandmother plants. For plants arising from non-pappus seeds, I did not find such a history influence.

Soil Feedback differences could not be correlated with differences in root infection with pathogenic fungi or colonization by AM fungi.

Conclusion: Consistent with the hypothesis, the results demonstrate that progeny from non-pappus seeds, which are produced for *in situ* persistence, are less negatively affected by conditions of 'self-cultivated' (trained) soil than progeny from pappus seeds, which easily disperse from the capitula and represent the long-distance dispersal type. Therefore, the magnitude of negative feedback corresponds to the dispersal potential of the different seed types giving an advantage to non-pappus progeny in the existing population. Plants grown from non-pappus seeds, moreover, showed proportionally greater investment in long-distance dispersal under unfavorable conditions of 'self-cultivated' soil, which may allow escape from sib competition and negative density effects.

5.1 Synthesis

Study I and II of this book highlight the importance of the ecological context in AM research. Recently, *A. artemisiifolia* has been repeatedly cited as an example of a plant whose invasive success is facilitated by AM fungi in the new range (e.g. Shah et al. 2009; Wurst et al. 2011; Grilli et al. 2012; Sanon et al. 2012; Emery and Rudgers 2012). This statement is based on a study by Fumanal et al. (2006) showing that *A. artemisiifolia* is colonized by AM fungi in different habitats in France and, moreover, demonstrating that single-grown plants produced more biomass when soil was inoculated with the AM fungus *Glomus intraradices*. In this regard, my research was motivated to find further evidence for *A. artemisiifolia* being promoted by AM fungi in the new European range under conditions of a maintained ecological context of the soil, natural AM fungal community and plant origin. However, when testing AM fungal communities from a roadside and a cornfield habitat, the presence of AM fungi increased performance of *A. artemisiifolia* in less fertile roadside soil only (Study I). Moreover, natural AM fungal communities did not increase the competitive ability of

A. artemisiifolia in the presence of co-existing mycorrhizal plant competitors in the new range (Study II). Furthermore, the positive effect of the roadside AM fungal community on *A. artemisiifolia* in isolation was reversed when *A. artemisiifolia* grew with a conspecific (Study II). Therefore, *A. artemisiifolia* is an unconvincing example of how an invasive plant is promoted by AM fungal associations in the new range. Contrary to Fumanal et al. (2006), the results of my studies, which are always based on experiments maintaining the ecological context of soil and AM fungi, show that the successful spread of *A. artemisiifolia* may not be related to the impact of AM fungi in Central Europe. These findings further illustrate that a plant species' response in isolation to an AM fungal isolate does not necessarily predict its response to natural AM fungal communities under comparatively high realism and its response to AM fungal communities in competitive situations.

Study III of this book demonstrates that the role of propagules in invasive plant processes can be ecologically very complex, in particular when invasive plants are heterocarpic, i.e., produce distinct types of seeds on a single plant. *G. parviflora* represents such a non-native, heterocarpic plant species, which produces pappus and non-pappus seeds. The findings of the Study III, both experimental and literature researched, illustrate the different strategies of the two seed types of *G. parviflora*: greater dispersal ability away from maternal plants of lighter pappus seeds is accompanied with shorter survival in the seed bank, and reduced performance of plants arising from pappus seeds in home soil, as well as lower competitive ability in mixed populations under moderate soil fertility compared to plants from non-pappus seeds, which experience less negative soil feedback. Therefore, heterocarpy in the non-native species *G. parviflora* may help that plant to successfully cope with disadvantageous conditions of 'self-cultivated' soil in existing populations (non-pappus seed type) and to colonize new habitats (pappus seed type).

5.2 Future perspectives

Further research into the successful spread of non-native *A. artemisiifolia* should focus on questions other than facilitation by AM fungi. The diametrically opposed results reported on the competiveness of *A. artemisiifolia* in the new range to date – competitively dominant to co-existing mycorrhizal plant species (Study II) and strongly competitively inferior to *Lolium multiflorum* (Leskovšek et al. 2012) – highlight the need for additional competition studies in different ecosystems. Evolutionary aspects, such as adaptation of *A. artemisiifolia* to soil conditions, as indicated by the present Study I for a roadside habitat, may also be interesting for further research and may give insights into rapid evolution of local adaptation of non-native plants in harsh environmental conditions of the new range. In this context, the spread of *A. artemisiifolia* along the road may also be studied from the perspective of adhesive seed transport by vehicles and propagule pressure (von der Lippe and Kowarik 2012).

Moreover, results presented in Study III strongly suggest further research into soil feedback of progeny from heterocarpic plants. It would be interesting to test responses of progeny from the different seed types of *G. parviflora* to different soil fractions, and to disentangle what kind of soil biota may have caused the greater negative feedback of plants grown from pappus compared to those from non-pappus seeds. Aside from *G. parviflora*, other annual plant species, like *Centaura solstitialis* L., *Crepis foetida* L., *Crepis sancta* (L.) Babc., *Hedypnois cretica* (L.) Dum. Cour., *Leontodon saxatilis* Lam. or *Picris echioides* L., which also produce seeds both equipped with a pappus and without, may represent appropriate study species. In addition, maternal history effects, i.e., influence of the seed type of mother/grandmother plants on heterocarpic progeny, need further evaluation and may be more common in heterocarpic plant species.

6 REFERENCES TO GENERAL INTRODUCTION AND SUMMARY

Allen MF (1991) The Ecology of Mycorrhizae. Cambridge Studies in Ecology series. Cambridge University Press, Cambridge.

Aschehoug ET, Metlen KL, Callaway RM, Newcombe G (2012) Fungal endophytes directly increase the competitive effects of an invasive forb. Ecology 93:3-8.

Augé RM (2001) Water relations, drought and vesicular-arbuscular mycorrhizal symbiosis. Mycorrhiza 11:3-42.

Augé RM, Sylvia DM, Park S, Buttery BR, Saxton AM, Moore JL, Cho KH (2004) Partitioning mycorrhizal influence on water relations of *Phaseolus vulgaris* into soil and plant components. Can J Bot 82:503-514.

Bais HP, Vepachedu R, Gilroy S, Callaway RM, Vivanco JM (2003) Allelopathy and exotic plants: from genes to invasion. Science 301:1377-1380.

Baker HG (1965) Characteristics and modes of origin of weeds. In: Baker HG, Stebbins GL (eds) The genetics of colonizing species, Academic Press, New York, pp. 147-172.

Baker HG (1974) The evolution of weeds. Annu Rev Ecol Syst 5:1-24.

Barney J, Whitlow T (2008) A unifying framework for biological invasions: the state factor model. Biol Invasions 10:259-272.

Blackburn TM, Pyšek P, Bacher S, Carlton JT, Duncan RP, Jarošík V, Wilson JRU, Richardson DM (2011) A proposed unified framework for biological invasions. Trends Ecol Evol 26:333-339.

Blossey B, Nötzold R (1995) Evolution of increased competitive ability in invasive nonindigenous plants – a hypothesis. J Ecol 83:887-889.

Blumenthal D, Mitchell CE, Pyšek P, Jarošík V (2009) Synergy between pathogen release and resource availability in plant invasion. PNAS 106:7899-7904.

Blumenthal DM (2006) Interactions between resource availability and enemy release in plant invasion. Ecol Lett 9:887-895.

Borowicz VA (2001) Do arbuscular mycorrhizal fungi alter plant-pathogen relations? Ecology 82:3057-3068.

Bossdorf O, Auge H, Lafuma L, Rogers WE, Siemann E, Prati D (2005) Phenotypic and genetic differentiation between native and introduced plant populations. Oecologia 144:1-11.

Bray SR, Kitajima K, Sylvia DM (2003) Mycorrhizae differentially alter growth, physiology, and competitive ability of an invasive shrub. Ecol Appl 13:565-574.

Brown JH, Sax DF (2004) An essay on some topics concerning invasive species. Austral Ecol 29:530-536.

Callaway RM, Bedmar EJ, Reinhart KO, Silvan CG, Klironomos J (2011) Effects of soil biota from different ranges on *Robinia* invasion: acquiring mutualists and escaping pathogens. Ecology 92:1027-1035.

Callaway RM, Ridenour WM (2004) Novel weapons: invasive success and the evolution of increased competitive ability. Front Ecol Environ 2:436-443.

Callaway RM, Thelen GC, Rodriguez A, Holben WE (2004a) Soil biota and exotic plant invasion. Nature 427:731-733.

Callaway RM, Thelen GC, Barth S, Ramsey PW, Gannon JE (2004b) Soil fungi interactions between the invader *Centaurea maculosa* and North American natives. Ecology 85:1062-1071.

Cassey P, Blackburn TM, Duncan RP, Chown SL (2005) Concerning invasive species: reply to Brown and Sax. Aust Ecol 30:475-480.

Catford J, Vesk P, Richardson DM, Pyšek P (2012) Quantifying invasion level: towards the objective classification of invaded and invasible ecosystems. Glob Change Biol 18:44-62.

Catford JA, Jansson R, Nilsson C (2009) Reducing redundancy in invasion ecology by integrating hypotheses into a single theoretical framework. Divers Distrib 15:22-40.

Cheptou PO, Carrue O, Rouifed S, Cantarel A (2008) Rapid evolution of seed dispersal in an urban environment in the weed *Crepis sancta*. PNAS 105:3796-3799.

CLIMAP Project Members (1976) The surface of the ice age earth. Science 191:1131-1136.

Colautti R, Grigorovich I, MacIsaac H (2006) Propagule pressure: a null model for biological invasions. Biol Invasions 8:1023-1037.

Colautti RI, MacIsaac HJ (2004) A neutral terminology to define 'invasive' species. Divers Distrib 10:135-141.

Colautti RI, Ricciardi A, Grigorovich IA, MacIsaac HJ (2004) Is invasion success explained by the enemy release hypothesis? Ecol Lett 7:721-733.

Colautti RI, Richardson DM (2009) Subjectivity and flexibility in invasion terminology: too much of a good thing? Biol Invasions 11:1225-1229.

Crawley MJ, Harvey PH, Purvis A (1996) Comparative ecology of the native and alien floras of the British Isles. Philos Trans R Soc B Biol Sci 351:1251-1259.

D'Antonio CM, Vitousek PM (1992) Biological invasions by exotic grasses, the grass/fire cycle, and global change. Ann Rev Ecol Syst 23:63-87.

Daehler CC (2001) Two ways to be an invader, but one is more suitable for ecology. Bull Ecol Soc Am 82:101-102.

Daehler CC (2003) Performance comparisons of co-occuring native and alien invasive plants: implications for conservation and restoration. Annu Rev Ecol Syst 34:183-211.

Davis MA (2011) Researching invasive species 50 years after Elton: a cautionary tale. In: Richardson DM (ed) Fifty years of invasion ecology: the legacy of Charles Elton, Wiley-Blackwell, Oxford, UK, pp. 269-276.

Davis MA, Grime JP, Thompson K (2000) Fluctuating resources in plant communities: a general theory of invasibility. J Ecol 88:528-534.

Davis MA, Thompson K (2000) Eight ways to be a colonizer; two ways to be an invader: a proposed nomenclature scheme for invasion ecology. Bull Ecol Soc Am 81:226-230.

Dawson W, Rohr RP, van Kleunen M, Fischer M (2012) Alien plant species with a wider global distribution are better able to capitalize on increased resource availability. New Phytol 194:859-867.

Diez JM, Dickie I, Edwards G, Hulme PE, Sullivan JJ, Duncan RP (2010) Negative soil feedbacks accumulate over time for non-native plant species. Ecol Lett 13:803-809.

Elton CS (1958) The Ecology of Invasions by Animals and Plants, Methuen, London, UK.

Emery SM, Rudgers JA (2012) Impact of competition and mycorrhizal fungi on growth of *Centaurea stoebe,* an invasive plant of sand dunes. Am Midl Nat 167:213-222.

Fumanal B, Plenchette C, Chauvel B, Bretagnolle F (2006) Which role can arbuscular mycorrhizal fungi play in the facilitation of *Ambrosia artemisiifolia* L. invasion in France? Mycorrhiza 17:25-35.

Gange AC, Brown VK, Aplin DM (2003) Multitrophic links between arbuscular mycorrhizal fungi and insect parasitoids. Ecol Lett 6:1051-1055.

Grilli G, Urcelay C, Galetto L (2012) Forest fragment size and nutrient availability: complex responses of mycorrhizal fungi in native–exotic hosts. Plant Ecol 213:155-165.

Hart MM, Reader RJ, Klironomos JN (2003) Plant coexistence mediated by arbuscular mycorrhizal fungi. Trends Ecol Evol 18:418-423.

Heger T, Trepl L (2003) Predicting biological invasions. Biol Invasions 5:313-321.

van der Heijden MGA, Klironomos JN, Ursic M, Moutoglis P, Streitwolf-Engel R, Boller T, Wiemken A, Sanders IR (1998) Mycorrhizal fungal diversity determines plant biodiversity, ecosystem variability and productivity. Nature 396:69-72.

Helgason T, Fitter AH (2009) Natural selection and the evolutionary ecology of the arbuscular mycorrhizal fungi (Phylum Glomeromycota). J Exp Bot 60:2465-2480.

Hierro JL, Maron JL, Callaway RM (2005) A biogeographical approach to plant invasions: the importance of studying exotics in their introduced and native range. J Ecol 93:5-15.

Hoeksema JD (2010) Ongoing coevolution in mycorrhizal interactions. New Phytol 187:286-300.

Huston MA (1979) A general hypothesis of species diversity. Am Nat 113:81-101.

Huston MA (2004) Management strategies for plant invasions: manipulating productivity, disturbance, and competition. Divers Distrib 10:167-178.

Imbert E, Escarré J, Lepart J (1997) Seed heteromorphism in *Crepis sancta* (Asteraceae): performance of two morphs in different environments. Oikos 79:325-332.

Inderjit (2005) Plant invasions: Habitat invasibility and dominance of invasive plant species. Plant Soil 277:1-5.

Inderjit, van der Putten WH (2010) Impacts of soil microbial communities on exotic plant invasions. Trends Ecol Evol 25:512-519.

Jeschke JM, Gómez Aparicio L, Haider S, Heger T, Lortie CJ, Pyšek P, Strayer DL (2012) Support for major hypotheses in invasion biology is uneven and declining. NeoBiota 14:1-20.

Ji B, Bentivenga SP, Casper BB (2010) Evidence for ecological matching of whole AM fungal communities to the local plant-soil environment. Ecology 91:3037-3046.

Johnson NC, Graham JH, Smith FA (1997) Functioning of mycorrhizal associations along the mutualism-parasitism continuum. New Phytol 135:575-585.

Johnson NC, Wilson GWT, Bowker MA, Wilson JA, Miller MR (2010) Resource limitation is a driver of local adaptation in mycorrhizal symbioses. PNAS 107:2093-2098.

Johnstone IM (1986) Plant invasion windows: a time-based classification of invasion potential. Biol Rev 61:369-394.

Joshi J, Vrieling K (2005) The enemy release and EICA hypothesis revisited: incorporating the fundamental difference between specialist and generalist herbivores. Ecol Lett 8:704-714.

Keane RM, Crawley MJ (2002) Exotic plant invasions and the enemy release hypothesis. Trends Ecol Evol 17:164-170.

Keller RP, Geist J, Jeschke JM, Kühn I (2011) Invasive species in Europe: ecology, status, and policy. Environmental Sciences Europe 23:23.

Kiers ET, Duhamel M, Beesetty Y, Mensah JA, Franken O, Verbruggen E, Fellbaum CR, Kowalchuk GA, Hart MM, Bago A, Palmer TM, West SA, Vandenkoornhuyse P, Jansa J, Bücking H (2011) Reciprocal rewards stabilize cooperation in the mycorrhizal symbiosis. Science 333:880-882.

van Kleunen M, Dawson W, Schlaepfer D, Jeschke JM, Fischer M (2010) Are invaders different? A conceptual framework of comparative approaches for assessing determinants of invasiveness. Ecol Lett 13:947-958.

Klironomos JN (2002) Feedback with soil biota contributes to plant rarity and invasiveness in communities. Nature 417:67-70.

Kowarik I (1995) Time lags in biological invasions with regard to the success and failure of alien species. In: Pyšek P, Prach K, Rejmánek M, Wade M (eds) Plant invasions – general aspects and special problems. Academic Publishing, Amsterdam, pp. 15-38.

Kowarik I (2002) Biologische Invasionen in Deutschland: zur Rolle nichteinheimischer Pflanzen. NeoBiota 1:5-24.

Kowarik I (2003) Biologische Invasionen: Neophyten und Neozoen in Mitteleuropa, Ulmer, Stuttgart, Germany.

Kowarik I, Pyšek P (2012) The first steps towards unifying concepts in invasion ecology were made one hundred years ago: revisiting the work of the Swiss botanist Albert Thellung. Divers Distrib 18:1243-1252.

Kowarik I, Starfinger U (2009) Neobiota: a European approach. NeoBiota 8:21-28.

Kühn I, Kowarik I, Kollmann J, Starfinger U, Bacher S, Blackburn TM, Bustamante RO, Celesti-Grapow L, Chytrý M, Colautti RI, Essl F, Foxcroft LC, García-Berthou E, Gollasch S, Hierro J, Hufbauer RA, Hulme PE, Jarošik V, Jeschke JM, Karrer G, Mack RN, Molofsky J, Murray BR, Nentwig W, Osborne B, Pyšek P, Rabitsch W, Rejmánek M, Roques A, Shaw R, Sol D, van Kleunen M, Vilà M, von der Lippe M, Wolfe LM, Penev L (2011) Open minded and open access: introducing NeoBiota, a new peer-reviewed journal of biological invasions. NeoBiota 9:1-12.

Kulmatiski A, Beard KH, Stevens JR, Cobbold SM (2008) Plant–soil feedbacks: a meta-analytical review. Ecol Lett 11:980-992.

Leskovšek R, Eler K, Batič F, Simončic A (2012) The influence of nitrogen, water and competition on the vegetative and reproductive growth of common ragweed (*Ambrosia artemisiifolia* L.). Plant Ecol 213:769-781.

Liao C, Peng R, Luo Y, Zhou X, Wu X, Fang C, Chen J, Li B (2008) Altered ecosystem carbon and nitrogen cycles by plant invasion: a meta-analysis. New Phytol 177:706-714.

von der Lippe M, Kowarik I (2012) Interactions between propagule pressure and seed traits shape human-mediated seed dispersal along roads. Perspect Plant Ecol 14:123-130.

Lodge DM (1993) Biological Invasions: Lessons for Ecology. Trends Ecol Evol 8:133-137.

Lonsdale WM (1999) Global patterns of plant invasions and the concept of invasibility. Ecology 80:1522-1536.

Mack RN, Simberloff D, Lonsdale WM, Evans H, Clout M, Bazzaz FA (2000) Biotic invasions: causes, epidemiology, global consequences, and control. Ecol Appl 10:689-710.

Mandák B (1997) Seed heteromorphism and plant life cycle: a review of literature. Preslia 69:129-159.

Mangla S, Inderjit, Callaway RM (2008) Exotic invasive plant accumulates native soil pathogens which inhibit native plants. J Ecol 96:58-67.

Marler MJ, Zabinski CA, Callaway RM (1999) Mycorrhizae indirectly enhance competitive effects of an invasive forb on a native bunchgrass. Ecology 80:1180-1186.

Melbourne BA, Cornell HV, Davies KF, Dugaw CJ, Elmendorf S, Freestone AL, Hall RJ, Harrison S, Hastings A, Holland M, Holyoak M, Lambrinos J, Moore K, Yokomizo H (2007) Invasion in a heterogeneous world: resistance, coexistence or hostile takeover? Ecol Lett 10:77-94.

Mitchell CE, Agrawal AA, Bever JD, Gilbert GS, Hufbauer RA, Klironomos JN, Maron JL, Morris WF, Parker IM, Power AG, Seabloom EW, Torchin ME, Vázquez DP (2006) Biotic interactions and plant invasions. Ecol Lett 9:726-740.

Mitchell CE, Blumenthal D, Jarošík V, Pulley EE, Pyšek P (2010) Controls on pathogen species richness in plants' introduced and native ranges: roles of residence time, range size, and host traits. Ecol Lett 13:1525-1535.

Mitchell CE, Power AG (2003) Release of invasive plants from fungal and viral pathogens. Nature 421:625-627.

Moles AT, Grubber MAM, Bonser SP (2008) A new framework for predicting invasive plant species. J Ecol 96:13-17.

Mooney H, Cleland E (2001) The evolutionary impact of invasive species. PNAS 98:5446-5451.

Moora M, Berger S, Davison J, Öpik M, Bommarco R, Bruelheide H, Kühn I, Kunin WE, Metsis M, Rortais A, Vanatoa A, Vanatoa E, Stout JC, Truusa M, Westphal C, Zobel M, Walther GR (2011) Alien plants associate with widespread generalist arbuscular mycorrhizal fungal taxa: evidence from a continental-scale study using massively parallel 454 sequencing. J Biogeogr 38:1305-1317.

Mühlenbach V (1979) Contributions to the Synanthropic (Adventive) Flora of the Railroads in St. Louis, Missouri, U.S.A. Ann Mo Bot Gard 66:1-108.

Mummey DL, Rillig MC (2006) The invasive plant species *Centaurea maculosa* alters arbuscular mycorrhizal fungal communities in the field. Plant Soil 288:81-90.

Munkvold L, Kjøller R, Vestberg M, Rosendahl S, Jakobsen I (2004) High functional diversity within species of arbuscular mycorrhizal fungi. New Phytol 164:357-364.

Newsham KK, Fitter AH, Watkinson AR (1995) Multi-functionality and biodiversity in arbuscular mycorrhizas. Trends Ecol Evol 10:407-411.

Nuñez MA, Horton TR, Simberloff D (2009) Lack of belowground mutualisms hinders *Pinaceae* invasions. Ecology 90:2352-2359.

Parker MA, Malek W, Parker IM (2006) Growth of an invasive legume is symbiont limited in newly occupied habitats. Divers Distrib 12:563-571.

Pejchar L, Mooney HA (2009) Invasive species, ecosystem services and human well-being. Trends Ecol Evol 24:497-504.

Pimentel D, Lach L, Zuniga R, Morrison D (2000) Environmental and economic costs of nonindigenous species in the United States. Bioscience 50:53-63.

Pimentel D, Zuniga R, Monison D (2005) Update on the environmental and economic costs associated with alien-invasive species in the United States. Ecol Econ 52:273-288.

Pringle A, Bever JD, Gardes M, Parrent JL, Rillig MC, Klironomos JN (2009) Mycorrhizal symbioses and plant invasions. Annu Rev Ecol Evol S 40:699-715.

van der Putten WH, Yeates GW, Duyts H, Reis CS, Karssen G (2005) Invasive plants and their escape from root herbivory: a worldwide comparison of the root-feeding nematode communities of the dune grass *Ammophila arenaria* in natural and introduced ranges. Biol Invasions 7:733-746.

Pyšek P, Richardson DM (2006) The biogeography of naturalization in alien plants. J Biogeogr 33:2040-2050.

Pyšek P, Richardson DM (2007) Traits associated with invasiveness in alien plants: Where do we stand? In: Nentwig W. (ed) Biological invasions, Ecological Studies 193, Springer-Verlag, Berlin & Heidelberg, pp. 97-126.

Pyšek P, Richardson DM, Rejmánek M, Webster GL, Williamson MK, Kirschner J (2004a) Alien plants in checklists and floras: towards better communication between taxonomists and ecologists. Taxon 53:131-143.

Pyšek P, Richardson DM, Williamson M (2004b) Predicting and explaining plant invasions through analysis of source area floras: some critical considerations. Divers Distrib 10:179-187.

Rai JPN, Tripathi RS (1987) Germination and plant survival and growth of *Galinsoga parviflora* Cav. as related to food and energy content of its ray- and disc-achenes. Acta Oecol 8:155-165.

Redecker D, Kodner R, Graham LE 2000 Glomalean fungi from the Ordovician. Science 289:1920-1921.

Reinhart KO, Callaway RM (2006) Soil biota and invasive plants. New Phytol 170:445-457.

Reinhart KO, Packer A, van der Putten WH, Clay K (2003) Plant-soil biota interactions and spatial distribution of black cherry in its native and invasive ranges. Ecol Lett 6:1046-1050.

Rejmánek M, Richardson DM (1996) What attributes make some plant species more invasive? Ecology 77:1655-1661.

Richardson DM, Pyšek P, Rejmanek M, Barbour MG, Panetta FD, West CJ (2000a) Naturalization and invasion of alien plants: concepts and definitions. Divers Distri 6:93-107.

Richardson DM, Allsopp N, D'Antonio CM, Milton SJ, Rejmánek M (2000b) Plant invasions – the role of mutualisms. Biol Rev Camb Philos 75:65-93.

Richardson DM, Pyšek P (2006) Plant invasions: merging the concepts of species invasiveness and community invasibility. Prog Phys Geog 30:409-431.

Richardson DM, Pyšek P (2008) Fifty years of invasion ecology – the legacy of Charles Elton. Divers Distrib 14:161-168.

Richardson DM, Williams PA, Hobbs RJ (1994) Pine invasions in the Southern Hemisphere: determinants of spread and invadability. J Biogeogr 21:511-527.

Rout M, Chrzanowski T (2009) The invasive *Sorghum halepense* harbors endophytic N2-fixing bacteria and alters soil biogeochemistry. Plant Soil 315:163-172.

Rout ME, Callaway RM (2009) An Invasive Plant Paradox. Science 324:734-735.

Rout ME, Callaway RM (2012) Interactions between exotic invasive plants and soil microbes in the rhizosphere suggest that 'everything is not everywhere'. Ann Bot London 110:213-222.

van Ruijven J, De Deyn GB, Berendse F (2003) Diversity reduces invasibility in experimental plant communities: the role of plant species. Ecol Lett 6:910-918.

Sakai AK, Allendorf FW, Holt JS, Lodge DM, Molofsky J, With KA, Baughman S, Cabin RJ, Cohen JE, Ellstrand NC, McCauley DE, O'Neil P, Parker IM, Thompson JN, Weller SG (2001) The population biology of invasive species. Annu Rev Ecol Syst 32:305-32.

Sala OE, Chapin FS, Armesto JJ, Berlow E, Bloomfield J, Dirzo R, Huber-Sanwald E, Huenneke LF, Jackson RB, Kinzig A, Leemans R, Lodge DM, Mooney HA, Oesterheld M, Poff NL, Sykes MT, Walker BH, Walker M, Wall DH (2000) Global biodiversity scenarios for the year 2100. Science 287:1770-1774.

Sanon A, Beguiristain T, Cébron A, Berthelin J, Sylla SN, Duponnois R (2012) Differences in nutrient availability and mycorrhizal infectivity in soils invaded by an exotic plant negatively influence the development of indigenous *Acacia* species. J Environ Manage 95:275-279.

Scheublin TR, van Logtestijn RSP, van der Heijden MAG (2007) Presence and identity of arbuscular mycorrhizal fungi influence competitive interactions between plant species. J Ecol 95:631-638.

Schroeder FG (1969) Zur Klassifizierung der Anthropochoren. Vegetatio 16:225-238.

Schüßler A, Schwarzott D, Walker C (2001) A new fungal phylum, the Glomeromycota: phylogeny and evolution. Mycol Res 105:1413-1421.

Seifert EK, Bever JD, Maron JL (2009) Evidence for the evolution of reduced mycorrhizal dependence during plant invasion. Ecology 90:1055-1062.

Shah MA, Reshi ZA, Khasa DP (2009) Arbuscular mycorrhizas: drivers or passengers of alien plant invasion. Bot Rev 75:397-417.

Shah MA, Reshi Z, Reshi I (2008) Mycorrhizosphere mediated mayweed chamomile invasion in the Kashmir Himalaya, India. Plant Soil 312:219-225.

Shea K, Chesson P (2002) Community ecology theory as a framework for biological invasions. Trends Ecol Evol 17:170-176.

Sher AA, Hyatt LA (1999) The disturbed resource-flux invasion matrix: a new framework for patterns of plant invasion. Biol Invasions 1:107-114.

Simberloff D, Von Holle B (1999) Positive interactions of nonindigenous species: invasional meltdown? Biol Invasions 1:21-32.

Smith SE, Read DJ (2008) Mycorrhizal Symbiosis, 3rd edn. Academic Press, London, UK.

Stinson KA, Campbell SA, Powell JR, Wolfe BE, Callaway RM, Thelen GC, Hallett SG, Prati D, Klironomos JN (2006) Invasive plant suppresses the growth of native tree seedlings by disrupting belowground mutualisms. PloS Biol 4:1-5.

Tanowitz BD, Salopek PF, Mahall BE (1987) Differential germination of ray and disc achenes in *Hemizonia increscens* (Asteraceae). Am J Bot 74:303-312.

Trocha LK, Kałucka I, Stasińska M, Nowak W, Dabert M, Leski T, Rudawska M, Oleksyn J (2012) Ectomycorrhizal fungal communities of native and non-native *Pinus* and *Quercus* species in a common garden of 35-year-old trees. Mycorrhiza 22:121-134.

Valéry L, Fritz H, Lefeuvre J-C, Simberloff D (2008) In search of a real definition of the biological invasion phenomenon itself. Biol Invasions 10:1345-1351.

Veresoglou SD, Rillig MC (2012) Suppression of fungal and nematode plant pathogens through arbuscular mycorrhizal fungi. Biology Lett 8:214-216.

Verhoeven KJF, Biere A, Harvey JA, van der Putten WH (2009) Plant invaders and their novel natural enemies: who is naïve? Ecol Lett 12:107-117.

Vermeij GJ (1991) When biotas meet - understanding biotic interchange. Science 253:1099-1104.

Vilà M, Basnou C, Pyšek P, Josefsson M, Genovesi P, Gollasch S, Nentwig W, Olenin S, Roques A, Roy D, Hulme PE, Andriopoulos P, Arianoutsou M, Augustin S, Bacher S, Bazos I, Bretagnolle F, Chiron F, Clergeau P, Cochard PO, Cocquempot C, Coeur d'Acier A, David M, Delipetrou P, Desprez-Loustau ML, Didžiulis V, Dorkeld F, Essl F, Galil BS, Gasquez J, Georghiou K, Gudžinskas Z, Hatzofe O, Hejda M, Jarošík V, Kark S, Kokkoris I, Kühn I, Lambdon PW, Lopez-Vaamonde C, Marcer A, Migeon A, McLoughlin M, Minchin D, Navajas M, Panov VE, Pascal M, Pergl J, Perglová I, Pino J, Poboljšaj K, Rabitsch W, Rasplus JY, Sauvard D, Scalera R, Sedláček O, Shirley S, Winter M, Yannitsaros A, Yart A, Zagatti P, Zikos A (2010) How well do we understand the impacts of alien species on ecosystem services? A pan-European, cross-taxa assessment. Front Ecol Environ 8:135-144.

Vitousek PM (1992) Global environmental change: an introduction. Annu Rev Ecol Syst 23:1-14.

Vitousek PM, Walker LR, Whiteaker LD, Mueller-Dombois D, Matson PA (1987) Biological invasion by *Myrica faya* alters ecosystem development in Hawaii. Science 238:802-804.

Vogelsang KM, Bever JD (2009) Mycorrhizal densities decline in association with nonnative plants and contribute to plant invasion. Ecology 90:399-407.

Wan F-H, Guo J-Y, Zhang F (2010) Booming Research on Biological Invasions in China. Aliens: The Invasive Species Bulletin 30:41-48.

Wardle DA, Bardgett RD, Callaway RM, van der Putten WH (2011) Terrestrial ecosystem responses to species gains and losses. Science 332:1273-1277.

Wardle DA, Bardgett RD, Klironomos JN, Setälä H, van der Putten WH, Wall DH (2004) Ecological linkages between aboveground and belowground biota. Science 304:1629-1633.

Webber BL, Scott JK (2012) Rapid global change: implications for defining natives and aliens. Global Ecol Biogeogr 21:305-311.

Wehner J, Antunes PM, Powell JR, Mazukatow J, Rillig MC (2010) Plant pathogen protection by mycorrhizas: diversity takes central role. Pedobiol 53:197-201.

Wilcove DS, Rothstein D, Dubow J, Phillips A, Losos E (1998) Quantifying threats to imperiled species in the United States. BioScience 48:607-615.

Williamson M (1996) Biological Invasions, Chapman & Hall, London, UK.

Williamson M, Brown KC (1986) The analysis and modelling of British invasions. Philos T Roy Soc B 314:505-522.

Williamson M, Fitter A (1996a) The varying success of invaders. Ecology 77:1661-1666.

Williamson MH, Fitter A (1996b) The characters of successful invaders. Biol Conserv 78:163-170.

Wolfe BE, Klironomos JN (2005) Breaking new ground: Soil communities and exotic plant invasion. Bioscience 55:477-487.

Wolfe BE, Rodgers VL, Stinson KA, Pringle A (2008) The invasive plant *Alliaria petiolata* (garlic mustard) inhibits ectomycorrhizal fungi in its introduced range. J Ecol 96:777-83.

Wurst S, Gebhardt K, Rillig MC (2011) Independent effects of arbuscular mycorrhiza and earthworms on plant diversity and newcomer plant establishment. J Veg Sci 22:1021-1030.

Young AM, Larson BMH (2011) Clarifying debates in invasion biology: A survey of invasion biologists. Environ Res 111:893-898.

APPENDIX A

The following table refers to hypotheses in invasion ecology, which are mentioned in the present work. The theories are ordered by three foci/categories that might be derived from hypotheses explaining the success of invasive plants in terrestrial ecosystems.

The categories are:
- i) features of the invasive species
- ii) characteristics of the new environment/habitat
- iii) interactions of invasive species with their new environment

Supplemental Table A.1 Overview on existing theories mentioned in the book.

Category: features of the invasive species

Hypothesis	Definition	References
Evolution of increased competitive ability (EICA)	Invasive success of individuals of a species in the new range is related to re-allocation of resources from defense mechanisms into growth and enhanced competitive abilities. (Individuals of a species taken from an area where they have been introduced produce more biomass than individuals taken from the species native range.) *Note:* EICA assumes absence of herbivores; hence, it is based on enemy release. First study species was *Lythrum salicaria* L. (purple loosestrife).	Blossey and Nötzold 1995; Bossdorf et al. 2005; Joshi and Vrieling 2005

Supplemental Table A.1 continued.

Category: features of the invasive species

Hypothesis	Definition	References
Ideal weed	Characteristics and life history traits of the invading species facilitate invasion by enabling them to outcompete native species. *Note:* Traits of an ideal weed are related to 'r-strategists' (ruderal life history, small seed size, high and early fecundity and fertility, rapid growth), as well as high phenotypic and genotypic plasticity.	Elton 1958; Baker 1965; Baker 1974; Rejmánek and Richardson 1996; Sakai et al. 2001; Pyšek and Richardson 2007
Lag phase	Many invasions are characterized by a lag phase followed by exponential range expansion. Species success may be delayed several decades or even centuries. *Note:* Factors associated with time lags might be intrinsic or extrinsic. Intrinsic factors are rate of population increases (e.g. late fecundity) or occurrence of evolutionary changes. Extrinsic factors are related to lacks of favorable environmental conditions ('invasion windows' (Johnstone (1986)).	Kowarik 1995

Supplemental Table A.1 continued.

Category: features of the invasive species

Hypothesis	Definition	References
Propagule pressure	High supply and frequency (number) of plant propagule introductions increase chance of successful invasion due to seed swamping, continual supplementation, high genetic diversity, greater probability of introduction to favorable environments, compensation of high mortality rates and bottleneck-situations, respectively. *Note:* Propagules are seeds, adult plants or reproductive vegetative fragments.	Williamson and Fitter 1996b; Lonsdale 1999; Pyšek and Richardson 2006; Colautti et al. 2006
Tens rule	The theory estimates the potential that a non-native species becomes invasive. It says that 10 % of imported species become introduced, 10 % of those introduced species become established, and 10 % of established species become a pest (i.e. cause high economic costs); hence, 1 in 10 of established species becomes highly invasive. *Note:* The theory is based on mathematical models regarding non-native species in Britain (both animals and plants). It refers to terms, which are related to environmental or economic impact (e.g. pest).	Williamson and Brown 1986; Williamson and Fitter 1996a

Supplemental Table A.1 continued.

Category: characteristics of the new environment

Hypothesis	Definition	References
Disturbance	Invasion success is related to disturbance events, which increase or decrease resource availability, which involves change in historical disturbance regimes. Non-native species have an equal chance of success at colonization and establishment compared to resident species because of changes in the rate or intensity of the turnover rate/flux of resources in a habitat. *Note:* Disturbance can be natural (e.g. floods, fires, hurricanes) or anthropogenic (e.g. introduction of cattle grazing, damming and straightening of rivers). Resources can include space, nutrients, or light. Disturbance alone is not always followed by invasion.	Sher and Hyatt 1999; Colautti et al. 2006
Dynamic equilibrium model	Invasion success is related to both disturbance and productivity. Non-native species can establish in low disturbance–low productivity systems, but only become dominant in high productivity systems with high levels of disturbance (required to establish). A change in disturbance regime can cause opposite effects in environments with contrasting levels of productivity.	Huston 2004

Supplemental Table A.1 continued.

Category: characteristics of the new environment		
Hypothesis	Definition	References
Dynamic equilibrium model (continued)	*Note:* The theory is based on the dynamic equilibrium model of species diversity (Huston 1979). Since disturbance can be both abiotic and biotic, increasing frequency and intensity of invasive species also alter disturbance regimes resulting in lower diversity of the community and dominance by the invasive species.	
Environmental heterogeneity	Predicts that heterogeneity both increases invasion success and reduces the impact to native species in the community, because it promotes invasion and coexistence mechanisms that would not possible in homogeneous environments. *Note:* Heterogeneity can result from abiotic heterogeneity, i.e. variation in the physical environment or biotic heterogeneity, i.e. variation in the occurrence and abundance of organisms. It can occur both at temporal and spatial scales (interaction neighborhood, local or regional metacommunity). Invasion processes depend on heterogeneity at local and regional metacommunity scales.	Melbourne et al. 2007

Supplemental Table A.1 continued.

Category: characteristics of the new environment

Hypothesis	Definition	References
Fluctuating resource availability	A plant community becomes more susceptible to invasion whenever there is an increase in the amount of unused resources. This can be due to increase in supply (e.g. abiotic disturbance such as rainfall) or a decrease in resource use (e.g. predation or die back of resident plants) or both. *Note:* The theory rests upon that an invading species must have access to available resources and that a species will have greater success in invading a community if it does not encounter intense competition for these resources from resident species.	Davis et al. 2000
Opportunity window	The ability of non-native species to successfully invade native communities is related to opportunities. These are niche opportunities, natural enemy escape opportunities and/or resource opportunities, which all vary in time and space. *Note:* The theory assumes that most species have periods of relative activity and relative inactivity during a year. Opportunities arise when resident species are relatively inactive and are not placing high demands on resources, or when invaders and residents differ in their responses to varying factors.	Shea and Chesson 2002

Supplemental Table A.1 continued.

Category: interactions between invasive species with their new environment

Hypothesis	Definition	References
Enemy release (ER)	Introduced to new ranges, non-native species experience a decrease in regulation by herbivores and other natural enemies, which results in a rapid increase in distribution and abundance. It assumes that generalists in the new range have greater impact on native than non-native species, and host switching events by specialist enemies of native congeners are rare. *Note:* Mitchell and Power (2003) studied 473 species and found support for escape from aboveground enemies: on average 84 % fewer fungi and 24 % fewer viruses infected plants in the introduced range in North America compared with their native range in Europe.	Keane and Crawley 2002; Mitchell and Power 2003; Colautti et al. 2004 (for arguments against ER)
Enhanced mutualisms	Non-native species experience greater positive effects from associations with mutualists, e.g. arbuscular mycorrhizas in the new range than in the native range. The impact of antagonists (pathogens) is similar in both ranges. The theory is based on the assumption that invasive species are able to associate with soil biota in the new range.	Reinhart and Callaway 2006

Supplemental Table A.1 continued.

Category: interactions between invasive species with their new environment

Hypothesis	Definition	References
Increased nitrogen cycling	The hypothesis assumes that non-native plants interact with soil microbes in a biogeographically explicit way resulting in greater nitrogen availability in the soil, from which invasive plants profit more than native plants. *Note:* The theory is based on the observation that invaded ecosystems show higher plant productivity and increased soil nitrogen pools and total ecosystem nitrogen stocks, respectively (Liao et al. 2008). Invasive plants often produce higher litter quality (contains a higher concentration of nitrogen).	Rout and Callaway 2009; Rout and Callaway 2012
Invasional meltdown	In invaded ecosystems, success of non-native species is enhanced by presence of other non-native species due to synergistic interactions among invaders ('invasion domino effect'). *Note:* Synergistic effects may arise from direct or indirect facilitative interactions that increase likelihood of survival and/or of ecological impact, and possibly the magnitude of impact of a non-native species.	Simberloff and Von Holle 1999

Supplemental Table A.1 continued.

Category: interactions between invasive species with their new environment		
Hypothesis	Definition	References
Novel weapons	Invasive plants profit from differences in species composition of competitors in the new range compared to their native range because invasive plants possess root exudates (allelochemicals), which are highly inhibitory to naïve plants in the new range. In their native range, these allelopathic compounds are relatively ineffective against natural neighbors due to co-evolutionary processes. *Note:* Famous example is *Centaurea stoebe* (former wrongly termed *Centaurea maculosa*), which releases (+/−)-Catechin; (−)-Catechin is phytotoxic.	Callaway and Ridenour 2004; Hierro et al. 2005; Bais et al. 2003
Mycorrhizal degradation	The theory says that disturbance events that disrupt mutualistic relationships such as mycorrhizas could facilitate the establishment of non-native species, which may become dominant, if they are less dependent on AM fungi and invest little in maintaining the soil community structure. The degraded soil community structure hinders successful re-establishment of native species. *Note:* It refers to ecosystems where native plants have strong mutualistic relationships with soil AM fungi. It assumes that i) increased disturbance intensity causes a decline in arbuscular mycorrhizas, ii) non-native species are often less dependent on mutualists (e.g. AM fungi), and iii) plants have different AM fungal hosting abilities.	Vogelsang et al. 2004

APPENDIX B
Supplemental Tables A.I.1–A.I.3 to Study I

Supplemental Table A.I.1 Analyses of variance (ANOVAs) on the second principal component score (PC2) of the Principal Component Analyses (PCAs) on plant biomass, root traits and mycorrhization of *Ambrosia artemisiifolia* in the experiment, with P.ori (plant origin), Soil, and Myc. (mycorrhizal treatment) as factors. Values in bold indicate significance at $P < 0.05$.

PC2 of PCA on		Plant Biomass		Root traits		Mycorrhization	
Factors	df	F	P	F	P	F	P
P.ori	1	0.18	0.672	**9.52**	**0.003**	0.57	0.453
Soil	1	0.20	0.654	3.90	0.053	0.22	0.638
Myc.	2	0.44	0.647	0.39	0.681	**148.71**	**<0.001**
P.ori x Soil	1	0.18	0.668	0.84	0.363	0.83	0.364
P.ori x Myc.	2	2.73	0.069	0.66	0.521	**3.29**	**0.040**
Soil x Myc.	2	0.75	0.473	0.80	0.455	**16.41**	**<0.001**
P.ori x Soil x Myc.	2	0.69	0.503	0.81	0.451	1.91	0.152
Residuals		132		54		132	

Supplemental Table A.I.2 Biomass traits, root traits and mycorrhization of *Ambrosia artemisiifolia* in response to soil and mycorrhizal treatment (Myc.Treat) in the experiment. Different lower case letters indicate significant differences ($P < 0.05$) among treatment groups according to Tukey's HSD test. Values represent mean ± SE.

	Roadside soil			Cornfield soil		
Myc.Treat	local	foreign	control	local	foreign	control
Biomass traits						
Shoot (mg)	98.6 ± 7.0[b]	106.3 ± 6.4[b]	81.6 ± 5.9[b]	304.4 ± 13.8[a]	350.6 ± 19.7[a]	349.8 ± 13.1[a]
Root (mg)	106.9 ± 11.4[b]	115.0 ± 11.2[b]	81.6 ± 9.1[b]	272.5 ± 14.8[a]	401.2 ± 47.1[a]	319.2 ± 19.5[a]
S5W	10 ± 2[b]	10 ± 2[b]	9 ± 2[b]	37 ± 6[a]	33 ± 9[a]	31 ± 5[a]
Total seeds	91 ± 9[b]	82 ± 6[bc]	63 ± 5[c]	174 ± 15[a]	173 ± 13[a]	215 ± 16[a]
WRS (mg)	43.9 ± 6.7[b]	43.9 ± 7.4[b]	45.4 ± 6.1[b]	88.1 ± 10.0[a]	108.2 ± 22[a]	88.1 ± 10.6[a]
WIS (mg)	49.8 ± 5.2[b]	51.8 ± 6.0[b]	32.4 ± 5.4[c]	93.1 ± 8.9[a]	90.9 ± 11.0[a]	123.9 ± 9.3[a]
WMI (mg)	26.8 ± 3.3[b]	32.2 ± 3.4[b]	27.6 ± 3.3[b]	71.6 ± 6.1[a]	78.2 ± 8.0[a]	82.8 ± 5.5[a]
Root traits						
RLV (km/m^3)	54.9 ± 5.1[c]	62.1 ± 6.5[bc]	47.3 ± 8.0[c]	86.7 ± 5.4[ab]	109.9 ± 6.3[a]	100.9 ± 4.7[a]
RSA (cm^2)	92.5 ± 8.6[c]	113.6 ± 13.0[c]	76.7 ± 14.1[c]	196.9 ± 15.3[b]	278.8 ± 27.5[a]	233.7 ± 13.7[ab]
FRV (mm^3)	9.1 ± 1.4[a]	8.1 ± 1.1[a]	8.8 ± 1.5[a]	10.2 ± 1.2[a]	12.6 ± 1.0[a]	10.6 ± 0.9[a]
FRL (m)	5.2 ± 0.9[a]	4.3 ± 0.7[a]	5.1 ± 0.9[a]	6.1 ± 0.8[a]	7.2 ± 0.7[a]	5.9 ± 0.6[a]

Supplemental Table A.I.2 continued.

	Roadside soil			Cornfield soil		
Myc.Treat	local	foreign	control	local	foreign	control
Root traits (continued)						
CRV (mm^3)	284.4 ± 30.8 [b]	399.4 ± 57.8 [b]	229.9 ± 48.7 [b]	1146.7 ± 121.7 [a]	2187.2 ± 518.8 [a]	1349.9 ± 133.7 [a]
CRL (m)	11.2 ± 1.1 [c]	14.3 ± 1.5 [bc]	9.0 ± 1.7 [c]	19.8 ± 1.6 [ab]	25.3 ± 1.5 [a]	24.2 ± 1.2 [a]
ARD (mm)	0.180 ± 0.005 [b]	0.192 ± 0.004 [b]	0.168 ± 0.006 [b]	0.239 ± 0.008 [a]	0.264 ± 0.012 [a]	0.245 ± 0.007 [a]
Mycorrhization						
TRC (%)	79 ± 3 [a]	76 ± 2 [a]	0 [c]	45 ± 3 [b]	49 ± 4 [b]	0 [c]
AMF.H (%)	21 ± 4 [c]	72 ± 2 [a]	0 [d]	40 ± 3 [b]	22 ± 3 [c]	0 [d]
AMF.A (%)	6 ± 1 [c]	28 ± 3 [a]	0 [d]	19 ± 2 [b]	7 ± 2 [c]	0 [d]
AMF.V (%)	2 ± 1 [b]	16 ± 2 [a]	0 [b]	2 ± 1 [b]	0.7 ± 0 [b]	0 [b]
FE.H (%)	69 ± 5 [a]	8 ± 2 [c]	0 [d]	5 ± 3 [cd]	32 ± 5 [b]	0 [d]
FE.A (%)	45 ± 4 [a]	5 ± 2 [c]	0 [c]	2 ± 1 [c]	22 ± 4 [b]	0 [c]
FE.V (%)	25 ± 3 [a]	3 ± 1 [c]	0 [c]	1 ± 1 [c]	9 ± 2 [b]	0 [c]

S5W, number of seeds produced after five weeks; Total seeds, total seed number; WRS, weight of ripe seeds; WIS, weight of immature seeds; WMI, weight of male inflorescences; RLV, root length per volume; RSA, root surface area; FRV, fine root volume; FRL, fine root length; CRV, coarse root volume; CRL, coarse root length; ARD, average root diameter; TRC, total AM fungal root colonization; AMF.H, colonization by coarse AM fungal hyphae; AMF.A, colonization by coarse AM fungal arbuscules; AMF.V, colonization by coarse AM fungal vesicles; FE.H, colonization by hyphae of fine endophytes; AMF.A, colonization by arbuscules of fine endophytes; AMF.V, colonization by vesicles of fine endophytes.

Supplemental Table A.I.3 Biomass variables and root traits of *Ambrosia artemisiifolia* in response to soil and plant origin of *A. artemisiifolia* in the experiment. Within a row, different lower case letters indicate significant differences ($P < 0.05$) among treatment groups according to Tukey's HSD test. Values represent mean ± SE.

Soil	roadside		cornfield	
Plant origin Variables	local	foreign	local	foreign
Shoot (mg)	96.3 ± 6.0 [b]	95.2 ± 5.0 [b]	339.3 ± 11.9 [a]	330.3 ± 14.8 [a]
Root (mg)	104.6 ± 9.1 [b]	98.2 ± 8.9 [b]	311.9 ± 17.8 [a]	349.8 ± 32.8 [a]
S5W	9 ± 1 [b]	10 ± 2 [b]	33 ± 6 [a]	35 ± 6 [a]
Total seeds	82 ± 5 [b]	75 ± 7 [b]	181 ± 12 [a]	193 ± 12 [a]
WRS (mg)	47.1 ± 5.5 [b]	41.6 ± 5.5 [b]	89.0 ± 11.5 [a]	100.6 ± 13.6 [a]
WIS (mg)	47.6 ± 4.1 [b]	42.0 ± 5.3 [b]	96.0 ± 6.8 [a]	108.5 ± 9.5 [a]
WMI (mg)	32.7 ± 2.7 [c]	25.0 ± 2.7 [c]	89.3 ± 6.0 [a]	66.0 ± 4.1 [b]
RLV (km/m^3)	53.2 ± 4.1 [b]	56.7 ± 7.2 [b]	90.6 ± 3.9 [a]	106.3 ± 5.1 [a]
RSA (cm^2)	87.9 ± 7.5 [b]	101.9 ± 13.5 [b]	216.8 ± 10.5 [a]	252.9 ± 21.1 [a]
FRV (mm^3)	9.2 ± 0.9 [a]	7.9 ± 1.3 [a]	8.7 ± 0.7 [a]	13.2 ± 0.7 [a]
FRL (m)	5.3 ± 0.6 [a]	4.3 ± 0.8 [a]	4.8 ± 0.4 [a]	7.8 ± 0.5 [a]
CRV (mm^3)	265.0 ± 28.3 [b]	352.0 ± 53.7 [b]	1286.2 ± 93.6 [a]	1790.5 ± 347.6 [a]
CRL (m)	10.6 ± 1.0 [b]	12.6 ± 1.5 [b]	22.3 ± 1.0 [a]	23.8 ± 1.5 [a]
ARD (mm)	0.173 ± 0.005 [b]	0.189 ± 0.003 [b]	0.253 ± 0.005 [a]	0.247 ± 0.010 [a]

S5W, number of seeds produced after five weeks; Total seeds, total seed number; WRS, weight of ripe seeds; WIS, weight of immature seeds; WMI, weight of male inflorescences; RLV, root length per volume; RSA, root surface area; FRV, fine root volume; FRL, fine root length; CRV, coarse root volume; CRL, coarse root length; ARD, average root diameter.

APPENDIX C

Supplemental Tables A.II.1–A.II.5 to Study II

Supplemental Table A.II.1 Vegetative and reproductive variables of the target–challenger experiment in response to relative density of *Ambrosia artemisiifolia* (grown as target or challenger), neighboring plant species (Neighbor species) and soil treatment (Soil treat.), i.e. AM fungi (AMF) or non-mycorrhizal control (control). Values represent mean ± SE.

Neighbor species	Soil treat.	*Ambrosia artemisiifolia*			Neighbor
		Shoot biomass (g)	No. female flowers	No. male flowers	Shoot biomass (g)
A. artemisiifolia grown as target					
Conyza	AMF	5.601 ± 0.318	0 ± 0	15.142 ± 2.394	0.191 ± 0.025
canadensis	control	4.345 ± 0.264	0 ± 0	4.571 ± 1.104	0.138 ± 0.013
Daucus	AMF	2.076 ± 0.190	0.143 ± 0.132	11.857 ± 3.047	0.546 ± 0.057
carota	control	3.036 ± 0.301	0.429 ± 0.397	8.714 ± 2.753	0.400 ± 0.042
Artemisia	AMF	2.703 ± 0.251	1.000 ± 0.670	3.429 ± 1.640	0.850 ± 0.108
vulgaris	control	2.685 ± 0.285	0.286 ± 0.171	6.714 ± 1.770	0.666 ± 0.068
Tanacetum	AMF	3.021 ± 0.267	0 ± 0	6.857 ± 2.094	0.641 ± 0.076
vulgare	control	2.004 ±0.338	0.571 ± 0.397	4.857 ±1.963	0.764 ±0.073
A. artemisiifolia grown as challenger					
Conyza	AMF	1.189 ± 0.048	1.643 ± 0.391	2.893 ± 0.394	0.073 ± 0.013
canadensis	control	1.101 ± 0.042	2.429 ± 0.683	2.536 ± 0.305	0.072 ± 0.007
Daucus	AMF	1.163 ± 0.051	1.679 ± 0.562	2.750 ± 0.295	0.159 ± 0.013
carota	control	1.179 ± 0.057	2.500 ± 0.539	3.714 ± 0.572	0.204 ± 0.017
Artemisia	AMF	1.125 ± 0.055	1.143 ± 0.281	3.000 ± 0.518	0.214 ± 0.033
vulgaris	control	1.235 ± 0.069	1.500 ± 0.492	2.107 ± 0.367	0.182 ± 0.045
Tanacetum	AMF	1.139 ± 0.071	0.929 ± 0.356	2.536 ± 0.491	0.280 ± 0.035
vulgare	control	1.134 ± 0.068	1.643 ± 0.410	1.857 ± 0.257	0.334 ± 0.110

Supplemental Table A.II.2 Response variables of the target–challenger experiment indicating a significant effect of neighboring species. Within a column, different lower case letters indicate significant differences ($P < 0.05$) among treatment groups according to Tukey's HSD test.

Neighboring plant species	Response variables		
	Shoot biomass of neighboring plant species (g)	Shoot biomass of *A. artemisiifolia* (g) in the presence of	Number of male flowers of *A. artemisiifolia* in the presence of
Conyza canadensis	0.118 ± 0.012 [c]	3.059 ± 0.386 [a]	6.300 ± 1.200 [a]
Daucus carota	0.327 ± 0.035 [b]	1.863 ± 0.172 [b]	6.800 ± 1.300 [a]
Artemisia vulgaris	0.478 ± 0.065 [a]	1.937 ± 0.173 [b]	3.813 ± 0.705 [a]
Tanacetum vulgare	0.496 ± 0.053 [a]	1.825 ± 0.184 [b]	4.027 ± 0.821 [a]

Supplemental Table A.II.3 Shoot biomass (mean ± SE) of *A. artemisiifolia* in response to its relative density (i.e. grown as target or challenger) and the four neighboring plant species. Different lower case letters indicate significant differences ($P < 0.05$) among treatment groups according to Tukey's HSD test.

Neighboring plant species	Shoot biomass of *A. artemisiifolia* (g) grown as	
	target	challenger
Conyza canadensis	4.973 ± 0.266 [a]	1.145 ± 0.034 [c]
Daucus carota	2.556 ± 0.219 [b]	1.171 ± 0.038 [c]
Artemisia vulgaris	2.694 ± 0.190 [b]	1.180 ± 0.046 [c]
Tanacetum vulgare	2.512 ± 0.255 [b]	1.137 ± 0.049 [c]

Supplemental Table A.II.4 Shoot biomass of *Ambrosia artemisiifolia* (mean ± SE) in response to neighboring plant species tested and soil treatment (AM fungi vs. non-mycorrhizal control) in the target–challenger experiment. Different lower case letters indicate significant differences ($P < 0.05$) among treatment groups according to Tukey's HSD test.

	Shoot biomass of *A. artemisiifolia* (g) grown in soil tratment	
Neighboring plant species	AM fungi	non-mycorrhizal
Conyza canadensis	3.395 ± 0.611 [a]	2.723 ± 0.454 [ab]
Daucus carota	1.619 ± 0.157 [c]	2.108 ± 0.292 [bc]
Artemisia vulgaris	1.914 ± 0.247 [bc]	1.960 ± 0.243 [bc]
Tanacetum vulgare	2.080 ± 0.287 [bc]	1.569 ± 0.208 [c]

Supplemental Table A.II.5 Percentages colonization by AM fungal structures (hyphae, arbuscules and vesicles) in roots of *Ambrosia artemisiifolia* and *Daucus carota* in the pairwise competition experiment with situations of intra- and interspecific competition.

	A. artemisiifolia		*D. carota*	
Competition	intraspecific	interspecific	intraspecific	interspecific
Hyphae (%)	23.2 ± 8.9	49.2 ± 9.4	47.5 ± 8.0	50.8 ± 10.4
Arbuscules (%)	15.0 ± 5.8	26.8 ± 7.6	30.9 ± 6.5	28.4 ± 7.0
Vesicles (%)	0.6 ± 0.4	8.6 ± 4.2	3.3 ± 1.2	11.2 ± 2.4

APPENDIX D

Supplemental Tables A.III.1–A.III.2 to Study III

Supplemental Table A.III.1 Mixed effect model analysis on percentages of AM fungal structures (hyphae, arbuscules, vesicles) and percentage root colonization by non-AM fungi in soils inoculated with trained soil. The mixed model was specified with Soil, Seed history, and Seed type as fixed effects. Plant genotype was treated as random effect. Values in bold indicate significance at $P < 0.05$.

Factors	Root colonization by hyphae P	AM arbuscules P	AM vesicles P	non-AM fungi P
Soil	0.222	**<0.001**	0.342	1.000
Seed history	0.083	0.121	0.752	0.362
Seed type	0.127	0.515	0.870	1.000
Soil x Seed history	1.000	0.685	0.600	0.520
Soil x Seed type	0.462	0.386	0.922	0.308
Seed history x Seed type	0.416	0.581	0.379	0.505
Soil x Seed history x Seed type	0.080	0.210	0.574	0.832

Supplemental Table A.III.2 Analysis of variance (ANOVA) on seed weight of seeds produced by *Galinsoga parviflora* during the second training round of the experiment, with Soil, Seed history, and Seed type as factors. Values in bold indicate significance at $P < 0.05$.

		Seed weight	
Factors	df	F	P
Soil	1	3.053	0.0939
Seed history	1	0.933	0.3441
Seed type	1	**38.272**	**<0.001**
Soil x Seed history	1	0.148	0.7043
Soil x Seed type	1	3.861	0.0616
Seed history x Seed type	1	0.227	0.6381
Soil x Seed history x Seed type	1	2.133	0.1577
Residuals	23		

i want morebooks!

Buy your books fast and straightforward online - at one of world's fastest growing online book stores! Environmentally sound due to Print-on-Demand technologies.

Buy your books online at
www.get-morebooks.com

Kaufen Sie Ihre Bücher schnell und unkompliziert online – auf einer der am schnellsten wachsenden Buchhandelsplattformen weltweit! Dank Print-On-Demand umwelt- und ressourcenschonend produziert.

Bücher schneller online kaufen
www.morebooks.de

VDM Verlagsservicegesellschaft mbH
Heinrich-Böcking-Str. 6-8   Telefon: +49 681 3720 174   info@vdm-vsg.de
D - 66121 Saarbrücken   Telefax: +49 681 3720 1749   www.vdm-vsg.de

Printed by Books on Demand GmbH, Norderstedt / Germany